Springer Theses

Recognizing Outstanding Ph.D. Research

Aims and Scope

The series "Springer Theses" brings together a selection of the very best Ph.D. theses from around the world and across the physical sciences. Nominated and endorsed by two recognized specialists, each published volume has been selected for its scientific excellence and the high impact of its contents for the pertinent field of research. For greater accessibility to non-specialists, the published versions include an extended introduction, as well as a foreword by the student's supervisor explaining the special relevance of the work for the field. As a whole, the series will provide a valuable resource both for newcomers to the research fields described, and for other scientists seeking detailed background information on special questions. Finally, it provides an accredited documentation of the valuable contributions made by today's younger generation of scientists.

Theses are accepted into the series by invited nomination only and must fulfill all of the following criteria

- They must be written in good English.
- The topic should fall within the confines of Chemistry, Physics, Earth Sciences, Engineering and related interdisciplinary fields such as Materials, Nanoscience, Chemical Engineering, Complex Systems and Biophysics.
- The work reported in the thesis must represent a significant scientific advance.
- If the thesis includes previously published material, permission to reproduce this must be gained from the respective copyright holder.
- They must have been examined and passed during the 12 months prior to nomination.
- Each thesis should include a foreword by the supervisor outlining the significance of its content.
- The theses should have a clearly defined structure including an introduction accessible to scientists not expert in that particular field.

More information about this series at http://www.springer.com/series/8790

Sabato Leo

CP Violation in $B_s^0 \to J/\psi\phi$ Decays

Measured with the Collider Detector at Fermilab

Doctoral Thesis accepted by
the University of Pisa, Italy

 Springer

Author
Dr. Sabato Leo
University of Illinois at Urbana-Champaign
Urbana, IL
USA

Supervisors
Prof. Giovanni Punzi
University of Pisa
Pisa
Italy

Dr. Diego Tonelli
CERN
Geneva
Switzerland

ISSN 2190-5053 ISSN 2190-5061 (electronic)
ISBN 978-3-319-36224-3 ISBN 978-3-319-07929-5 (eBook)
DOI 10.1007/978-3-319-07929-5

Springer Cham Heidelberg New York Dordrecht London

Supervisors' Foreword

Particle physics is now experiencing exciting and puzzling times. Recently, a massive spin-0 particle compatible with the Higgs boson predicted by the standard model (SM) was discovered. In addition, null results from experimental searches excluded non-SM particles as heavy as 1 TeV. These striking findings conclusively establish the standard model as an accurate and consistent description of the fundamental constituents of matter and their interactions at energies extending up to the "Fermi scale" (100 GeV), at which electromagnetic and weak interactions act as an unified force—and beyond.

However, a number of experimental indications and solid theoretical arguments support the idea that the standard model becomes necessarily incomplete at some energy higher than the 1 TeV scale directly probed thus far, and should be rooted in a more general theory. The description of such a theory, and the search for its experimental signatures, is the chief goal of today's fundamental physics.

Direct observation of feebly interacting massive particles requires accelerators capable of producing collisions with multi-MHz rates, at energies a few times larger than the mass of the sought particles. This is a technologically and economically daunting task that requires ever-larger international collaboration enterprises, enduring for decades.

The study of weak interactions of quarks, or *quark-flavor physics*, offers an attractive complementary path, since such interactions are sensitive to energy scales far higher than those directly accessible with current and foreseen colliders. The presence of non-SM dynamics can be indirectly, but unambiguously, inferred from quark transitions that allow exchanges of virtual non-SM particles. This requires focusing on processes that are experimentally accessible, sensitive to a broad class of SM extensions, and whose phenomenology is accurately predictable. Discrepancies emerging from the comparison of accurate SM predictions with equally precise experimental results would reveal the presence of non-SM dynamics.

Indeed, the discovery potential of this approach is no novelty. Indirect searches have been instrumental in directing us toward discovery of some of the most fundamental building blocks of the standard model itself. For instance, the existence of the charmed quark was conjectured and its mass estimated, a few years prior to its direct discovery, based on the observation of an anomalous suppression in K meson decay rates.

In two decades of extensive explorations, no search in quark-flavor revealed the presence of non-SM physics. Nevertheless, sub-10 % contributions from non-SM dynamics are not ruled out, especially in processes involving transitions of bottom-strange mesons (B_s^0 and $\bar{B}_s^0$). These bound states of a bottom and a strange quark offer a rich phenomenology that, at the time of this work, was only marginally explored. The decays of B_s^0 and $\bar{B}_s^0$ mesons into the final state involving a J/ψ and a ϕ meson stand out as one of the few *golden* channels that offer the most promising discovery opportunities. Interference between direct decays and decays following particle–antiparticle transitions, $B_s^0 \leftrightarrow \bar{B}_s^0 \to J/\psi\phi$, provides experimental access to the underlying mixing phase, a parameter that is very sensitive to a broad class of non-SM processes, and challenging to measure otherwise. Experimentally, a sizable decay rate into fully reconstructed final states, with intermediate resonances decaying into a muon pair and two charged particles makes the accurate reconstruction of large event samples accessible in hadron collisions and aids effective suppression of backgrounds. The precision measurement of the B_s^0 mixing phase is one of the key goals of current experimental particle physics.

This thesis reports the final measurement of the mixing phase performed in proton–antiproton collisions at center-of-mass energy of 1.96 TeV, provided by the Tevatron collider and recorded by the Collider Detector at Fermilab experiment (CDF). Sifting through 10^{14} collisions from 10 years of operations, a low-background sample of about 11,000 $B_s^0, \bar{B}_s^0 \to J/\psi\phi$ decays is reconstructed. A sophisticated analysis of the data is then performed, which encompasses in a single measurement all most advanced experimental techniques employed in flavor-physics at hadron colliders. In collisions that typically yield tenths to hundreds of charged particles, the distance traveled by the B_s^0 or $\bar{B}_s^0$ meson before decaying is measured with 30 μm precision. This, combined with a few permil measurement of the meson's momentum, provides a measurement of decay time with about 90 fs resolution. A suite of algorithms, specifically optimized for this thesis, is used to study the particle activity in the whole event searching for tiny correlations between particle species and their electrical charges, which allow determining whether the bottom-strange particle was produced as a meson or an antimeson. In addition, spatial correlations between the directions of final-state particles are exploited in determining the value of the orbital angular momentum between decay products. All these informations are encapsulated into a technically challenging multivariate likelihood fit with more than 30 parameters to determine the mixing phase.

The relevance and impact of the results is well worth the effort. The determination of the mixing phase and several supplementary quantities are among the most precise available from a single experiment. They have been published in a highly cited letter to Physical Review, PRL 109, 171802 (2012), where this thesis is also cited. No indication of deviations from the SM predictions is found and significantly improved bounds on the allowed phenomenology of non-SM particles or interactions are obtained.

The experimental exploration of the bottom-strange mixing phase is rapidly progressing owing to abundant samples available at the Large Hadron Collider, and promises many further new insights. But this thesis will remain a valuable resource, providing a clear and detailed description of a legacy measurement performed by the CDF experiment, where measurements of the bottom-strange mixing phase were first performed, and many of the key techniques that have now become standard in hadron-collisions flavor physics had been pioneered and perfected.

Pisa, 2014

Prof. Giovanni Punzi
Dr. Diego Tonelli

Abstract

We report on the final measurement of the *CP*-violating B_s^0 mixing phase $\beta_s^{J/\psi\phi}$ *in* $\sqrt{s} = 1.96$ TeV proton-antiproton collisions collected with the Collider Detector at the Fermilab Tevatron collider. Using a sample corresponding to 9.6 fb^{-1} of integrated luminosity, we fit the decay-time evolution of $B_s^0 \to J/\psi(\to \mu^+\mu^-)\,\phi\,(\to K^+K^-)$ decays in which the *b*-quark content at production and the *CP* parity of the final state are identified. The interference of decays with and without mixing renders the B_s^0 mixing phase observable. The phase is determined to be $-0.06 < \beta_s^{J/\psi\phi} < 0.30$ at 68 % confidence level. The decay-width difference between heavy and light B_s^0 eigenstates, and the average B_s^0 lifetime are also determined to be $\Delta\Gamma_s = 0.068 \pm 0.027$ ps^{-1} and $\tau_s = 1.528 \pm 0.021$ ps, respectively. These results are among the world's most precise from a single experiment, and compatible with standard model predictions.

Contents

Chapter 1
Introduction

Over the past three decades, the standard model (SM) of elementary particle physics has been established by many experimental verifications as the basic theory describing fundamental particles and their interactions at energy scales from $\mathcal{O}(1)$ eV to $\mathcal{O}(1)$ TeV. However, serious theoretical open questions, and some puzzling experimental results lead to the common prejudice that the SM is an effective theory valid at the energies probed so far, which should be completed and complemented by a more general framework at higher energies. General theoretically well-founded arguments indicate that new particles or interactions could be relevant at energy scales of $\mathcal{O}(1)$ TeV and above. The search for such extensions of the SM has become a chief goal in today's particle physics. The diverse and numerous activities in this program are broadly classified in two different but complementary approaches.

Direct searches use the highest-energy collisions attainable to directly produce on-shell non-SM particles and identify their presence through their decay products or interactions. The interpretation of the results obtained is usually straightforward, but the reach is limited by the maximum energies achievable.

Indirect searches aim at inferring the presence of new particles or interactions in *low energy* processes (rare processes, flavor oscillations, etc.) by seeking discrepancies between experimental measurements and theory predictions. Indirect searches are typically sensitive to higher energy scales [1, 2] than attainable directly, and have the potential to reveal richer information on the underlying dynamics. However, the interpretation of results is often complicated by irreducible uncertainties or assumptions in the theory predictions, and the measurements are often more challenging because aiming at high precision.

The flavor physics of quarks is considered one of the most promising frameworks to pursue such indirects searches. The quark-flavor sector of the SM accounts for 10 out of 18 free SM parameters, linked with some of the most profound unanswered questions on the SM, such as those on the origin of the observed intriguing pattern of quark masses and interaction hierarchies. The abundance of accessible physics processes makes the study of quark transitions promising to gain insight on possible realizations of SM extensions, especially through the phenomenology of the

© Springer International Publishing Switzerland 2015
S. Leo, *CP Violation in $B_s^0 \to J/\psi\phi$ Decays*, Springer Theses,
DOI 10.1007/978-3-319-07929-5_1

violation of charge-parity symmetry (*CP*). The violation of *CP* symmetry is the non-invariance of physics processes when all spatial coordinates are inverted and particles are replaced by their antiparticles. Its study is considered one of the most promising gateways to indicate the presence of non-SM particles or interactions. Dedicated kaon experiments and "*B*-factories" have provided very stringent constraints on the presence of non-SM physics in leading (and some subleading) processes involving charged and neutral kaons and bottom hadrons [3, 4], allowing the redundant determination of precise quark-mixing parameters of the Cabibbo-Kobayashi-Maskawa (CKM) matrix [5, 6]. All measurements performed so far are consistently described at first order by the CKM picture [3]. Nevertheless, subleading effects from non-SM physics are not yet ruled out [7].

The dynamics of bottom-strange mesons (B_s^0) specifically, is a promising field of investigation since a comparatively smaller amount of experimental information is available [4].

The study of B_s^0 dynamics provides privileged access to the V_{ts} element of CKM matrix, which remains experimentally poorly constrained by the results on B^0, B^+, and K meson dynamics and is sensitive to a broad class of plausible SM extensions. This is the chief motivation for pursuing B_s^0 physics. Because of its peculiar phenomenology, B_s^0 meson dynamics is best studied at high-energy hadron colliders. The Collider Detector at Fermilab (CDF), has led the study of B_s^0 dynamics over the course of the past decade through reconstruction of large samples of B_s^0 decays in $\sqrt{s} = 1.96\,\text{TeV}$ proton-antiproton collisions.

In this thesis we report the final search for non-SM physics in B_s^0 mixing using $B_s^0 \rightarrow J/\psi(\rightarrow \mu^+\mu^-)\phi(\rightarrow K^+K^-)$ decays at CDF. CDF is a large multipurpose solenoidal magnetic spectrometer, surrounded by 4π projective calorimeters and fine-grained muon detectors.

The CKM hierarchy predicts very small *CP* violation in $B_s^0 \rightarrow J/\psi\phi$ decays with a good theoretical precision. But *CP* violation in these processes can be significantly increased in a broad class of SM extensions. This feature makes $B_s^0 \rightarrow J/\psi\phi$ a *golden channel* in the indirect search for new physics [8, 9]. Specifically, the time-dependent *CP* asymmetry of the decays allows probing the phase $\beta_s^{J/\psi\phi} = \arg[V_{ts}V_{tb}^\star/(V_{cs}V_{cb}^\star)]$, which is closely related to the phase of the $B_s^0 - \bar{B}_s^0$ oscillation amplitude. The overall constraints of the CKM-matrix yield $\beta_s^{J/\psi\phi} \approx 0.02$ [10], which is negligible with respect to the current experimental sensitivity. Any significant deviation in the observed value would unambiguously indicate presence of non-SM physics.

The first measurement of the *CP*-violating phase in $B_s^0 \rightarrow J/\psi\phi$ decays was finalized in 2008 by the CDF experiment [11]. It showed a mild, 1.5σ discrepancy from the SM. Similar and consistent effects were found by the D0 experiment [12], yielding a combination that deviated 2.2σ from the SM [13]. This attracted some interest, further enhanced by independent results on a correlated observable from the D0 collaboration [14, 15]. Later measurements of $\beta_s^{J/\psi\phi}$ have pointed towards a better consistency with the CKM expectations [16–19]. However, the understanding of the $B_s^0 - \bar{B}_s^0$ mixing phenomenology is still far from being finalized [10], which calls for more precise and redundant measurements.

This final measurement of $\beta_s^{J/\psi\phi}$ supersedes the previously-published CDF result using half of the current data set [16]. In addition to the mixing phase, we report measurements of the decay-width difference between the light and heavy mass eigenstates of the B_s^0 meson, $\Delta\Gamma_s$; of their mean lifetime, τ_s; and of the angular-momentum composition of $B_s^0 \to J/\psi\phi$ decays.

I conducted all the steps of the data analysis, from the event-selection and simulation to the final determination of the systematic uncertainties. My specific original contribution with respect to previous versions of the measurement has been developing a significantly improved understanding of the algorithms that identify the b quark content of the strange-bottom meson at production. I demonstrated that a more refined selection of the control samples of data than typically adopted results in a more precise determination of tagging performances, simplifying the analysis and reducing the systematic uncertainties.

The results presented in this thesis are among the most precise determinations available to date from a single experiment and agree with measurements from other experiments and with the SM expectations [10, 17–20]. They contribute to establish that large non-SM contributions to the $B_s^0 - \bar{B}_s^0$ mixing phase are unlikely, and provide important constraining information for phenomenological model building. All results are published in a letter to *Physical Review* [21].

The first chapter of the thesis is dedicated to the motivation of the measurement and the detailed description of the angular and decay-time distribution of the $B_s^0 \to J/\psi\phi$ decays, along with a short review of the current experimental situation and an outline of the analysis. Chapter 3 describes the experimental apparatus, with a brief description of the Tevatron accelerator, the CDF II detector, and the data-taking operations. The reconstruction and selection of the data set is described in Chap. 4. Chapter 5 focuses mainly on experimental acceptance and algorithms used to determine the flavor of the B_s^0 meson at production, and their calibrations. The calibration of all the others analysis tools needed for the measurement are also reported in this chapter. The maximum likelihood estimator is discussed in Chap. 6 along with the tests aimed at studying its properties. Chapter 7 discuss the systematic effects. The results are reported in Chap. 8. A few final pages are devoted to draw conclusions and discuss future prospects.

References

1. G. Isidori et al., Flavor physics constraints for physics beyond the standard model. Ann. Rev. Nucl. Part. Sci. **60**, 355 (2010) arXiv:1002.0900 [hep-ph]
2. M. Ciuchini, A. Stocchi, Physics opportunities at the next generation of precision flavor physics. Ann. Rev. Nucl. Part. Sci. **61**, 491 (2011) arXiv:1110.3920 [hep-ph]
3. J. Beringer et al. (Particle Data Group), Review of particle physics. RPP. Phys. Rev. **D86**, 010001 (2012)
4. Y. Amhis et al. (Heavy Flavor Averaging Group), Averages of b-hadron, c-hadron, and tau-lepton properties as of early 2012 (2012) arXiv:1207.1158 [hep-ex]. And updates at http://www.slac.stanford.edu/xorg/hfag

5. N. Cabibbo, Unitary symmetry and leptonic decays. Phys. Rev. Lett. **10**, 531 (1963)
6. M. Kobayashi, T. Maskawa, CP violation in the renormalizable theory of weak interaction. Prog. Theor. Phys. **49**, 652 (1973)
7. J. Charles et al. (CKMfitter Group), CP violation and the CKM matrix: assessing the impact of the asymmetric B factories. Eur. Phys. J. **C41**, 1 (2005) hep-ph/0406184. Updated results and plots available at http://ckmfitter.in2p3.fr
8. I. Dunietz et al., In pursuit of new physics with B_s decays. Phys. Rev. D. **63**, 114015 (2001) arXiv:hep-ph/0012219 [hep-ph]
9. A. Datta et al., New physics in $b \to s$ transitions and the $B_{d,s}^0 \to V_1 V_2$ angular analysis. Phys. Rev. **D86**, 076011 (2012) arXiv:1207.4495 [hep-ph]
10. A. Lenz, Theoretical update of B-mixing and lifetimes (2012) arXiv:1205.1444 [hep-ph]
11. T. Aaltonen et al. (CDF Collaboration), First flavor-tagged determination of bounds on mixing-induced CP violation in $\bar{B}_s^0 \to J/\psi\phi$ decays. Phys. Rev. Lett. **100**, 161802 (2008) arXiv:0712.2397 [hep-ex]
12. V. Abazov et al. (D0 Collaboration), Measurement of B_s^0 mixing parameters from the flavor-tagged decay $B_s^0 \to J/\psi\phi$. Phys. Rev. Lett. **101**, 241801 (2008) arXiv:0802.2255 [hep-ex]
13. The CDF and D0 Collaborations, Combination of D0 and CDF results on $\Delta\Gamma_s$ and the CP-violating phase $\beta_s^{J/\psi\phi}$, CDF Note 9787, D0 Note 5928-CONF (2009)
14. V. M. Abazov et al. (D0 Collaboration), Evidence for an anomalous like-sign dimuon charge asymmetry. Phys. Rev. Lett. **105**, 081801 (2010) arXiv:1007.0395 [hep-ex]
15. V.M. Abazov et al. (D0 Collaboration), Measurement of the anomalous like-sign dimuon charge asymmetry with 9 fb^{-1} of $p\bar{p}$ collisions. Phys. Rev. **D84**, 052007 (2011) arXiv:1106.6308 [hep-ex]
16. T. Aaltonen et al. (CDF Collaboration), Measurement of the CP-violating phase $\beta_s^{J/\psi\phi}$ in $\bar{B}_s^0 \to J/\psi\phi$ decays with the CDF II detector. Phys. Rev. **D85**, 072002 (2012) arXiv:1112.1726 [hep-ex]
17. V. Abazov et al. (D0 Collaboration), Measurement of the CP-violating phase $\phi_s^{J/\psi\phi}$ using the flavor-tagged decay $\bar{B}_s^0 \to J/\psi\phi$ in 8 fb^{-1} of $p\bar{p}$ collisions. Phys. Rev. **D85**, 032006 (2012) arXiv:1109.3166 [hep-ex]
18. R. Aaij et al. (LHCb Collaboration), Measurement of the CP-violating phase ϕ_s in the decay $\bar{B}_s^0 \to J/\psi\phi$. Phys. Rev. Lett. **108**, 101803 (2012) arXiv:1112.3183 [hep-ex]
19. G. Aad et al. (ATLAS Collaboration), Time-dependent angular analysis of the decay $\bar{B}_s^0 \to J/\psi\phi$ and extraction of $\Delta\Gamma_s$ and the CP-violating weak phase ϕ_s by ATLAS (2012) arXiv:1208.0572 [hep-ex]
20. R. Aaij et al. (LHCb Collaboration), Measurement of the polarization amplitudes and triple product asymmetries in the $B_s^0 \to \phi\phi$ decay. Phys. Lett. **B713**, 369 (2012) arXiv:1204.2813 [hep-ex]
21. T. Aaltonen et al. (CDF Collaboration), Measurement of the bottom-strange meson mixing phase in the full CDF data set. Phys. Rev. Lett. **109**, 171802 (2012) arXiv:1208.2967 [hep-ex]

Chapter 2
Flavor as a Probe for Non-SM Physics

This chapter introduces the motivations for the measurement described in this thesis. After a brief description of the general theoretical framework we focus on CP violation in $B_s^0 - \bar{B}_s^0$ mixing through $B_s^0 \rightarrow J/\psi\phi$ decays as probe for non-standard model physics. An overview of the measurement, and a summary of the current experimental status are also presented.

2.1 The Current Landscape

The standard model (SM) of particle physics provides a quantum field theoretical description of three fundamental interactions, namely the strong, weak, and electromagnetic interactions, that act among the elementary spin-half particles, the quarks and the leptons [1]. The SM structure is based on symmetries of the Lagrangian for transformations of a gauge group, resulting in interactions being mediated by spin-one force carriers: eight massless gluons for the strong interaction; two charged massive bosons, $W^{\pm}$, and a single neutral massive boson, Z^0, for the weak interaction; and a massless photon, γ, for the electromagnetic interaction. Finally, the SM includes a spin-zero particle, the Higgs boson, which is the scalar excitation of the field that provides generation of particles masses through the spontaneous symmetry breaking of the gauge group of the electroweak interaction. Quarks and leptons interact via Higgs-mediated interactions that, unlike gauge interactions, are not ruled by symmetry principles. These *Yukawa* interactions are responsible for *flavor* physics. The term flavor is used to differentiate among the variety of species of quarks and leptons that have same quantum charges: up-type quarks (u, c, t), down-type quarks (d, s, b), charged leptons (e, μ, τ), and neutrinos (ν_e, ν_μ, ν_τ), each featuring three flavors.

In July 2012, the CMS and ATLAS Collaborations at the Large Hadron Collider (LHC) announced the discovery of a resonance produced in proton–proton collisions [2, 3]. The new particle has a mass of approximately $125\,\text{GeV/c}^2$, with

© Springer International Publishing Switzerland 2015

S. Leo, *CP Violation in $B_s^0 \rightarrow J/\psi\phi$ Decays*, Springer Theses,

DOI 10.1007/978-3-319-07929-5_2

properties compatible with the SM Higgs interpretation. The discovery of the Higgs boson completes the validation of the SM, as an extremely predictive theory capable of accurately explaining most of the experimental phenomena probed so far [4]: we know today that the physics of fundamental particles and interactions at energies in the sub-eV-TeV range is successfully described by this theory.

However, a number of solid theoretical arguments, and some experimental results motivate a prejudice that the SM should be a low-energy restriction of a more general theory that includes additional particles and interaction couplings. For instance, classical gravity, well described by general relativity, should break down at energy scales close to 10^{19} GeV, the Planck scale, at which quantum effects of gravity should become relevant [5]. The SM would be necessarily invalidated at such energy calling for a theory of quantum gravity to integrate the SM at the Planck-scale energies. In addition, the calculation of the Higgs-boson mass is affected by divergences due to radiative corrections that invalidate the SM at an energy scale that depends on the mass itself. This energy scale represents the *cut-off* of the effective model, i.e., the energy above which the model needs likely to be extended by a more fundamental theory. Either a fine-tuning of the model parameters that can push the cut-off at the Planck scale, or non-SM particles whose virtual contributions eliminate the divergence were proposed [6]. The observed value of the Higgs-boson mass consolidates the SM at the electroweak scale, and moves the cut-off at larger energy [4]. Hence, the question if non-SM particles are present in the energy range from $\mathcal{O}(1)$ TeV to the Planck scale is open, motivated also by cosmological arguments based on a large mismatch between the quantity of baryonic and luminous matter in the universe and astrophysical observations [7].

2.2 Flavor as a Probe of Non-SM Physics

Non-SM particles can be produced *directly* in high-energy collisions and observed through their decay products, provided that the available center-of-mass energy is sufficient to produce the heavy particles with sensible rates. The heavier are particles that can be produced, the higher the physics scale probed. The reach of this *direct* approach crucially depends on the unknown energy scale of the non-SM physics particles. Pushing forward the energy frontier requires devising technologies to achieve ever-higher center-of-mass energies. However, the non-SM particles can become directly observable at energies not reachable with current and foreseen technology.

A complementary approach is to infer the presence of non-SM particles *indirectly* in processes where they could be virtually exchanged between SM particles, by detecting deviations of observables from expectations precisely calculated in the SM [8, 9]. In indirect searches, the production threshold energy is not as critical as in direct searches. Because quantum effects become smaller the heavier are the virtual particles at play, higher non-SM physics scales are explored by increasing the precision of the measurements while controlling the SM contributions with sufficient accuracy to identify unanbiguosly non-SM effects.

The flavor physics of quarks is among the most promising sector for indirect searches. Experimental access to a plethora of precisely measurable processes, along with a mature phenomenology that provides many accurate predictions, allows the redundant determination of several SM parameters that can be compared for precision tests of the overall picture [10]. Indeed, flavor physics has proved very successful in building the current understanding of particle physics. For instance, the theoretical *ansatz* to explain [11] the suppressed decay rate of quark transitions that change the strangeness flavor by two units (such as $K^0 \rightarrow \mu^+\mu^-$ decays) was crucial to postulate the existence of a then-unknown charm quark and estimate its mass, before its experimental discovery [12, 13]. Similarly, an important prediction of the large value of the top-quark mass before its direct observation [14, 15], was inferred from the indirect constraints imposed by the measurements of $B^0 - \bar{B}^0$ mesons oscillations [16].

The physics of flavor is the physics of matter at its most fundamental level. It consists in the study of underlying patterns in the family replications of quarks and leptons, and in the highly hierarchical structure of their masses and couplings. The flavor sector of the SM accounts for 10 out 18 free parameters of the theory and still is impressively predictive and peculiar. Flavor violation is allowed only in the quark sector. Weak interactions mediated by $W^\pm$ bosons that change flavor of quarks (flavor-changing charged-currents, FCCC) are universal; flavor transitions mediated by neutral currents (flavor-changing neutral-currents, FCNC) are highly suppressed. The last cannot occur at *tree-level*, i.e., through the mediation of a W boson only, but they do require the intermediate exchange of a quark and a W boson (*loop* transition). In Fig. 2.1 we show this features with two examples of Feynman diagrams representing the two flavor-changing transitions in terms of the elementary particles involved. The FCNC are further suppressed in the SM by the Glashow-Illiopoulos-Maiani (GIM) mechanism [11], namely the smallness of the mass differences between second- and first-generation quarks, and by the hierarchical structure of *quark-mixing* angles, which determines the rotation of the quark-flavor basis with respect to the

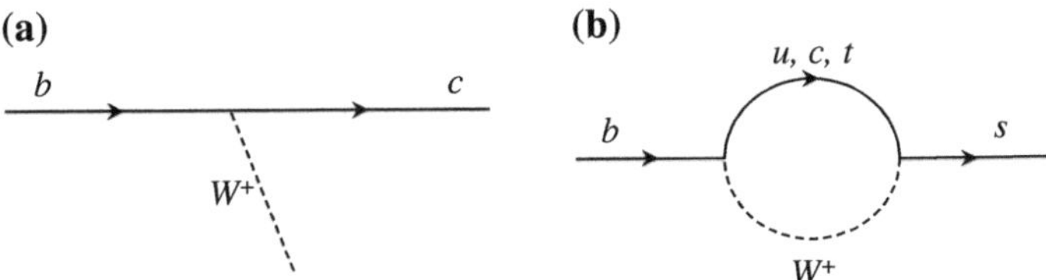

Fig. 2.1 Feynman graphs of two examples of flavor transition of the b quark. In **a**, the *tree*-level transition $b \rightarrow c$, where the b quark changes flavor and charge $(-1/3)$ becoming a c quark (with charge $+2/3$) through emission of a W^+ boson; in **b**, the *loop*-mediated transition $b \rightarrow s$, where the b quark changes its flavor by exchanging an up-type quark (either u, or c, or t) and a W boson with the s quark. In this case, the charge of the initial quark and the charge of the final quark are the same. Tree-level transition involves quarks of different type (up-down and down-up), while loop-mediated transitions can change the flavor of two quarks of the same type (up-up and down-down)

weak-interaction basis. These phenomenological features determine the observed pattern of quark transitions, and any extension of the SM must account for them.

A generic effective-theory approach [8] is powerful for describing non-SM physics effects in flavor physics, in a model-independent way. Assuming the non-SM physics scale to be higher than the electroweak energy scale, non-SM physics effects can be described by a generalization of the Fermi theory. In this approach, the SM Lagrangian is included in a more general local Lagrangian, which includes a series of operators with dimension $d > 4$, $O_i^{(d)}$, constructed in terms of SM fields, with arbitrary couplings $c_i^{(d)}$ suppressed by inverse powers of an effective scale Λ, which represents the cut-off of the effective theory:

$$\mathcal{L}_{\text{eff}} = \mathcal{L}_{\text{SM}} + \sum_i \frac{c_i^{(d)}}{\Lambda^{(d-4)}} O_i^{(d)} \tag{2.1}$$

Based on naturalness principle, bounds on Λ can be derived assuming an effective coupling $c_i \approx 1$; alternatively, bounds on the respective couplings can be determined assuming that $\Lambda \approx \mathcal{O}(1)\,\text{TeV}$. This approach allows useful and unified interpretation of many experimental results to derive stringent constraints on extensions of the SM in terms of few parameters (the coefficients of the higher-dimensional operators).

For instance, in a generic non-SM physics model, where suppression of FCNC processes is due only to the large masses of the particles that mediate them, i.e., the couplings are of order one, bounds on the scale Λ that are compatible with the measurements of FCNC decay rates are determined. Depending on the process under study, this approach yields bounds of order $\Lambda \gtrsim 10^2\,\text{TeV}$. Hence, either non-SM degrees of freedom emerge at energies higher than the TeV scale, or any SM extention at TeV scale must have a highly non-generic flavor structure, i.e., the coupling c_i should have very suppressed values. Table 2.1 lists the bounds derived from measurements related to the mixing of neutral mesons for a choice of operators that change the flavor of the decaying quark by two unit in Eq. (2.1), showing the predictive power of this indirect approach in establishing general features of the theory (either its energy scale or its flavor structure), which hold independently of the dynamical details of the model.

The phenomenology of the B, D, and K mesons is particularly useful for this purpose [10]. This thesis presents the analysis of the $B_s^0 \to J/\psi\phi$ decay, which

Table 2.1 Bounds from experimental constraints on meson-mixing [8]

Bounds on Λ (TeV)	Bounds on c_i ($\Lambda = 1$ TeV)	Mesons
10^2–10^5	10^{-11}–10^{-7}	$K^0 - \bar{K}^0$
10^3–10^4	10^{-7}–10^{-8}	$D^0 - \bar{D}^0$
10^2–10^3	10^{-7}–10^{-6}	$B^0 - \bar{B}^0$
10^2–10^3	10^{-5}	$B_s^0 - \bar{B}_s^0$

allows a measurement of the $B_s^0 - \bar{B}_s^0$ mixing phase, an extremely powerful and still largely unconstrained experimental probe for a large class of non-SM physics phenomena.

2.3 CKM Matrix and *CP*-violation

In the SM the only source of flavor-changing interactions is originated from a rotation of the quarks flavor basis with respect to the weak-interaction basis in the Yukawa sector. Such rotation is given by a unitary 3×3 complex matrix,

$$V_{\mathrm{CKM}} = \begin{pmatrix} V_{ud} & V_{us} & V_{ub} \\ V_{cd} & V_{cs} & V_{cb} \\ V_{td} & V_{ts} & V_{tb} \end{pmatrix},$$

known as the Cabibbo-Kobayashi-Maskawa (CKM) quark-mixing matrix [17, 18]. The constraints of unitarity of the CKM-matrix on the diagonal terms implies that the sum of all couplings of any of the up-type quarks to all the down-type quarks is the same for all generations, named *weak universality*, and derives from all $SU(2)$ doublets coupling with same strength to the vector bosons of weak interactions. Thus FCCC occurs at tree-level, while FCNC are mediated only by loops. A unitary $n \times n$ matrix contains n^2 independent real parameters, $2n - 1$ of those can be eliminated by rephasing the n up-type and n down-type fermion fields (changing all fermions by the same phase obviously does not affect V_{CKM}); hence, there are $(n-1)^2$ physical parameters left. A unitary matrix is also orthogonal, and as such it contains $n(n-1)/2$ parameters corresponding to the independent rotation angles between the n basis vectors; thus the remaining $(n-1)(n-2)/2$ parameters must be complex phases. For $n = 2$, i.e. two families, only one mixing angle remains, the Cabibbo angle and, no complex phases [17]. For $n = 3$ there are four physical parameters, namely three Euler angles and one irriducible phase, which provides a gateway for *CP* violation. The unitarity of the CKM matrix, $V_{\mathrm{CKM}} V_{\mathrm{CKM}}^{\dagger} = 1$, leads to 9 equations,

$$\sum_{k \in \{u,c,t\}} V_{ki} V_{kj}^{\star} = \delta_{ij} \qquad (i, j \in \{d, s, b\}).$$

Six require the sum of three complex quantities to vanish, and define triangles in the complex plane. The area of each triangle equals $J_{CP}/2$. The symbol J_{CP} identifies the Jarlskog invariant [19], whose size quantifies the magnitude of violation of *CP* symmetry in the SM. The *CP* symmetry is violated only if $J_{CP} \neq 0$, as confirmed by current measurements [10]: $J_{CP} = (2.884^{+0.253}_{-0.053}) 10^{-5}$. Any *CP*-violating quantity in the SM is proportional to J_{CP}, reflecting the fact that a single complex phase appears in the 3×3 CKM matrix. This feature makes the SM implementation of *CP* violation predictive, because all possible *CP* asymmetry measurements are correlated by their common origin from a single parameter of the theory.

The current knowledge of the CKM matrix elements magnitudes assuming unitarity is as follows [10]:

$$|V_{\text{CKM}}| = \begin{pmatrix} 0.97426^{+0.00022}_{-0.00014} & 0.22539^{+0.00062}_{-0.00095} & 0.003501^{+0.000196}_{-0.000087} \\ 0.22526^{+0.00062}_{-0.00095} & 0.97345^{+0.00022}_{-0.00018} & 0.04070^{+0.00116}_{-0.00059} \\ 0.00846^{+0.00043}_{-0.00015} & 0.03996^{+0.00114}_{-0.00062} & 0.999165^{+0.000024}_{-0.000048} \end{pmatrix}. \tag{2.2}$$

The observed hierarchy $|V_{ub}| \ll |V_{cb}| \ll |V_{us}|$, and $|V_{cd}| \ll 1$, suggests an expansion in powers of $\lambda = |V_{us}| \approx 0.23$, the sine of the Cabibbo angle [20, 21]

$$V_{\text{CKM}} = \begin{pmatrix} 1 - \lambda^2/2 - \lambda^4/8 & \lambda & A\lambda^3(\rho - i\eta) \\ -\lambda + A^2\lambda^5[1 - 2(\rho + i\eta)]/2 & 1 - \lambda^2/2 - \lambda^4(1 + 4A^2)/8 & A\lambda^2 \\ A\lambda^3[1 - (1 - \lambda^2/2)(\rho + i\eta)] & -A\lambda^2 + A\lambda^4[1 - 2(\rho + i\eta)]/2 & 1 - A^2\lambda^4/2 \end{pmatrix} + \mathcal{O}(\lambda^6), \tag{2.3}$$

where $A \approx 0.80$, $\rho \approx 0.14$ and $\eta \approx 0.34$ are real parameters [10].

One triangular equation of particular phenomenological interest is referred to as the unitarity triangle (UT), because all three terms are roughly of the same size,

$$V_{ud} V_{ub}^{\star} + V_{cd} V_{cb}^{\star} + V_{td} V_{tb}^{\star} = 0; \tag{2.4}$$

The UT equation is normalized as

$$R_t e^{-i\beta} + R_u e^{i\gamma} = 1, \tag{2.5}$$

where

$$R_t = \left| \frac{V_{td} V_{tb}^{\star}}{V_{cd} V_{cb}^{\star}} \right|, \quad R_u = \left| \frac{V_{ud} V_{ub}^{\star}}{V_{cd} V_{cb}^{\star}} \right|, \quad \beta = \arg\left(-\frac{V_{cd} V_{cb}^{\star}}{V_{td} V_{tb}^{\star}} \right), \quad \gamma = \arg\left(-\frac{V_{ud} V_{ub}^{\star}}{V_{cd} V_{cb}^{\star}} \right), \tag{2.6}$$

are, respectively, two sides and two angles of the UT. The third side is the unit vector, and the third angle is $\alpha = \pi - \beta - \gamma = \arg[-V_{td} V_{tb}^{\star} / V_{ud} V_{ub}^{\star}]$. Equation (2.5) shows that all the information related to the UT is encoded in one complex number,

$$\bar{\rho} + i\bar{\eta} = R_u e^{i\gamma}, \tag{2.7}$$

which corresponds to the coordinates $(\bar{\rho}, \bar{\eta})$ of the only nontrivial apex of the UT in the complex plane. Assuming that flavor-changing processes are fully described by the SM, the consistency of the various measurements with this assumption can be verified. The values of λ and A are known accurately from $K \to \pi l \nu$ and $b \to c l \nu$ decays respectively [7] to be

$$\lambda = 0.2257 \pm 0.0010, \quad A = 0.814 \pm 0.022. \tag{2.8}$$

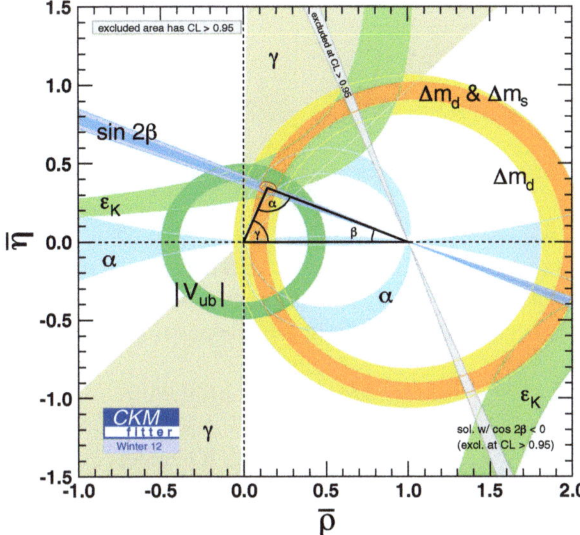

Fig. 2.2 Constraints in the $(\bar{\rho}, \bar{\eta})$ plane. The *red hashed region* of the global combination corresponds to 68 % CL

All the relevant observables are then expressed as a function of the two remaining parameters $\bar{\rho}$ and $\bar{\eta}$, and checks are performed on whether there exists a range in the $(\bar{\rho}, \bar{\eta})$ plane that is consistent with all measurements. The resulting constraints in the $(\bar{\rho}, \bar{\eta})$ plane are shown in Fig. 2.2. The overall consistency is impressive, yielding the following values for $\bar{\rho}$ and $\bar{\eta}$ [7]:

$$\bar{\rho} = +0.135^{+0.031}_{-0.016}, \quad \bar{\eta} = +0.349 \pm 0.017. \tag{2.9}$$

This support the ansatz that flavor and *CP* violation in flavor-changing processes are dominated by the CKM mechanism. Such remarkable success of the SM suggests that arbitrary non-SM physics contributions in flavor-changing processes that occurs at tree-level are highly suppressed with respect to SM contributions [8].

A greater chance for detecting the effects of non-SM physics might reside in the study of loop-mediated FCNC transitions, such as the ones that mediates neutral-mesons oscillations described in Sect. 2.3.1.

2.3.1 B_s^0 Oscillations

Because of flavor mixing, flavored neutral mesons are subject to particle-antiparticle oscillations through weak transitions that change the flavor by two units, $\Delta F = 2$. The $\Delta B = 2$ FCNC quark transition that drives the $B_s^0 - \bar{B}_s^0$ oscillations is depicted

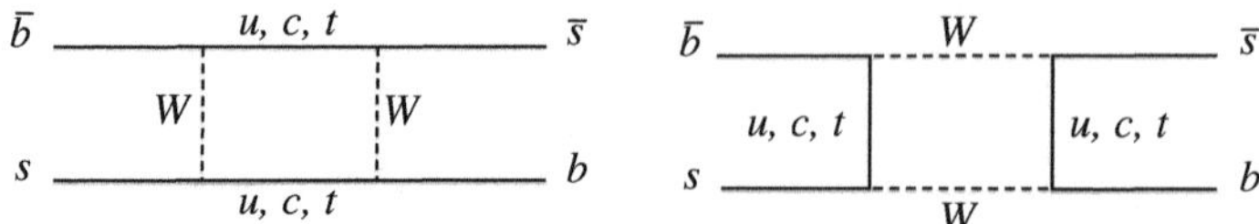

Fig. 2.3 Feynman diagrams of transitions associated with $B_s^0 - \bar{B}_s^0$ oscillations

in Feynman diagrams of Fig. 2.3, called *box* diagrams. As a result of flavor mixing, a pure B_s^0 or $\bar{B}_s^0$ state at time $t = 0$, such as the meson states created from $p\bar{p} \to b\bar{b}$ interactions at the Tevatron, evolves to be a superposition of B_s^0 and $\bar{B}_s^0$ at time t:

$$|\psi(t)\rangle = a(t)|B_s^0\rangle + b(t)|\bar{B}_s^0\rangle. \tag{2.10}$$

In the Wigner-Weisskopf approximation, which neglects corrections to the exponential decay rate at very low or very high times, the effective Hamiltonian of the system can be written as

$$\mathbf{H_q} = \begin{pmatrix} M_{11}^s & M_{12}^s \\ M_{12}^{s*} & M_{22}^s \end{pmatrix} - \frac{i}{2} \begin{pmatrix} \Gamma_{11}^s & \Gamma_{12}^s \\ \Gamma_{12}^{s*} & \Gamma_{22}^s \end{pmatrix}, \tag{2.11}$$

where $M_{11}^s = M_{22}^s$ and $\Gamma_{11}^s = \Gamma_{22}^s$ hold under the assumption of *CPT* invariance which is a fairly general assumption thus far confirmed by any experimental verification [7]. The off-diagonal elements M_{12}^s and Γ_{12}^s are responsible for B_s^0–$\bar{B}_s^0$ mixing phenomena. The *dispersive* part M_{12}^s corresponds to virtual $\Delta B = 2$ transitions dominated by heavy internal particles (top quarks in the SM) while the *absorptive* part Γ_{12}^s arises from on-shell transitions due to decay modes common to B_s^0 and $\bar{B}_s^0$ mesons. Diagonalization of the Hamiltonian leads to two mass eigenstates $B_{H,L}^s$ (H and L denote heavy and light, respectively), with mass $M_{H,L}^s$ and decay width $\Gamma_{H,L}^s$. The mass eigenstates are linear combinations of flavor eigenstates with complex coefficients p and q that satisfy $|p|^2 + |q|^2 = 1$,

$$|B_{L,H}^s\rangle = p|B_s^0\rangle \pm q|\bar{B}_s^0\rangle. \tag{2.12}$$

The time evolution of the mass eigenstates is governed by the two eigenvalues, $m_H - \frac{i}{2}\Gamma_H$ and $m_L - \frac{i}{2}\Gamma_L$,

$$|B_{H,L}^s(t)\rangle = e^{-i(m_{H,L}+i\Gamma_{H,L}/2)t}|B_{H,L}^s(0)\rangle. \tag{2.13}$$

The time evolution of B_s^0 and $\bar{B}_s^0$ is derived from Eq. (2.13) and their definition in Eq. (2.12),

$$|B_s^0(t)\rangle = g_+(t)|B_s^0\rangle + \frac{q}{p}g_-(t)|\bar{B}_s^0\rangle,$$

$$|\bar{B}_s^0(t)\rangle = g_+(t)|\bar{B}_s^0\rangle + \frac{p}{q}g_-(t)|B_s^0\rangle, \tag{2.14}$$

where

$$g_+(t) = e^{-imt-\frac{\Gamma}{2}t}\left[\cosh\frac{\Delta\Gamma t}{4}\cos\frac{\Delta mt}{2} - i\sinh\frac{\Delta\Gamma t}{4}\sin\frac{\Delta mt}{2}\right],$$

$$g_-(t) = e^{-imt-\frac{\Gamma}{2}t}\left[-\sinh\frac{\Delta\Gamma t}{4}\cos\frac{\Delta mt}{2} + i\cosh\frac{\Delta\Gamma t}{4}\sin\frac{\Delta mt}{2}\right], \tag{2.15}$$

which satisfy

$$|g_\pm(t)|^2 = \frac{e^{-\Gamma t}}{2}\left[\cosh\frac{\Delta\Gamma t}{2} \pm \cos\Delta mt\right],$$

$$g_+^\star(t)g_-(t) = -\frac{e^{-\Gamma t}}{2}\left[\sinh\frac{\Delta\Gamma t}{2} + i\sin\Delta mt\right], \tag{2.16}$$

where $m = (m_H+m_L)/2$, $\Gamma = (\Gamma_H+\Gamma_L)/2$, $\Delta m = m_H-m_L$ and $\Delta\Gamma = \Gamma_L-\Gamma_H$.

The probabilities of observing a B_s^0 at any time t if the meson was produced as either a B_s^0 or a $\bar{B}_s^0$ at $t = 0$ are

$$\mathcal{P}(B_s^0 \to B_s^0) = |\langle B_s^0(t)|B_s^0(0)\rangle|^2 \tag{2.17}$$

$$= |g_+(t)|^2 = \frac{e^{-\Gamma t}}{2}\left[\cosh\frac{\Delta\Gamma t}{2} + \cos\Delta mt\right],$$

$$\mathcal{P}(\bar{B}_s^0 \to B_s^0) = |\langle B_s^0(t)|\bar{B}_s^0(0)\rangle|^2$$

$$= \left|\frac{p}{q}\right|^2 |g_-(t)|^2 = \left|\frac{p}{q}\right|^2 \frac{e^{-\Gamma t}}{2}\left[\cosh\frac{\Delta\Gamma t}{2} - \cos\Delta mt\right].$$

The values of M_{12}^s and Γ_{12}^s are physical observables and can be determined from measurements of the following quantities (for more details see, e.g., Ref. [22]):

- the mass difference between the heavy and light mass eigenstates

$$\Delta m_s \equiv m_H^s - m_L^s \approx 2|M_{12}^s|\left(1 - \frac{|\Gamma_{12}^s|^2}{8|M_{12}^s|^2}\sin^2\phi_{12}^s\right), \tag{2.18}$$

where $\phi_{12}^s = \arg(-M_{12}^s/\Gamma_{12}^s)$ is convention-independent;
- the decay width difference between the light and heavy mass eigenstates

$$\Delta\Gamma_s \equiv \Gamma_L^s - \Gamma_H^s \approx 2|\Gamma_{12}^s|\cos\phi_{12}^s\left(1 + \frac{|\Gamma_{12}^s|^2}{8|M_{12}^s|^2}\sin^2\phi_{12}^s\right); \tag{2.19}$$

- the flavor-specific asymmetry

$$a_{\text{sl}}^s \equiv \frac{|p/q|^2 - |q/p|^2}{|p/q|^2 + |q/p|^2} \approx \frac{|\Gamma_{12}^s|}{|M_{12}^s|} \sin \phi_{12}^s \approx \frac{\Delta\Gamma_s}{\Delta m_s} \tan \phi_{12}^s . \tag{2.20}$$

The correction terms proportional to $\sin^2 \phi_{12}^s$ in Eqs. (2.18) and (2.19) are irrelevant compared to the present size of experimental uncertainties. In addition, the ratio of q and p can be expressed as

$$\left(\frac{q}{p}\right) = -\frac{\Delta m_s + \frac{i}{2}\Delta\Gamma_s}{2(M_{12}^s - \frac{i}{2}\Gamma_{12}^s)}. \tag{2.21}$$

The possibility of flavor oscillations strongly enriches the phenomenology of CP violation, which occurs in B_s^0 meson decays through three different manifestations. Considering a decay in a CP eigenstate f with eigenvalue η_f and A_f being the decay amplitude of $B_s^0 \to f$, we define the following classes of CP violation:

- *CP violation in decay* or *direct CP violation*, which is the only possible CP violating effect in charged meson decays since they cannot undergo mixing, occurs when the amplitude of decay to a final state is not the same as the amplitude of the CP conjugate of the initial state decaying to the CP conjugate of the final state, $|\bar{A}_{\bar{f}}|/|A_f| \neq 1$. In the $B_s^0 \to J/\psi\phi$ channel, the standard model CP-violating weak phase in the decay is suppressed by a factor of λ^2 [23]. Hence, the assumption of no direct CP violation in $B_s^0 \to J/\psi\phi$ decays, $|\bar{A}_{\bar{f}}| = |A_f|$, holds to a very good approximation.
- *CP violation in mixing* occurs when $|q/p| \neq 1$. In the B_s^0 meson system, the CKM model predicts $|q/p| = 1 + \mathcal{O}(10^{-3})$ [24]. In semileptonic B_s^0 decays this leads to a charge asymmetry in the decay products, but in $B_s^0 \to J/\psi\phi$ the factor $|q/p|$ is not isolated, therefore CP violation in mixing is not directly measured in this analysis.
- *CP violation due to interference between decays with and without mixing* may appear in the evolution of B_s^0 and $\bar{B}_s^0$ mesons decay. This type of CP violation is observable by measuring the phase difference between the amplitude for a direct decay to a final state f and the amplitude for a decay produced by oscillation, discussed in the next section.

2.4 Analysis of the Time-Evolution of $B_s^0 \to J/\psi\phi$ Decays

For decays dominated by the $b \to c\bar{c}s$ tree amplitude, the phase difference is denoted by

$$\phi_s \equiv -\arg\left(\eta_f \frac{q}{p}\frac{\bar{A}_f}{A_f}\right), \tag{2.22}$$

where A_f and $\bar{A}_f$ are the decay amplitudes of $B_s^0 \to f$ and $\bar{B}_s^0 \to f$, respectively. In the absence of direct CP violation $\bar{A}_f/A_f = \eta_f$. With these approximations, the CP violating phases appearing in B_s^0 mixing reduce to the phase $\phi_s \approx -2\beta_s^{J/\psi\phi}$, defined as [25]

$$\beta_s^{J/\psi\phi} \equiv \arg\left(-\frac{V_{ts}V_{tb}^*}{V_{cs}V_{cb}^*}\right). \tag{2.23}$$

If non-SM physics occurs in M_{12}^s or in the decay amplitudes, the measured value of $\beta_s^{J/\psi\phi}$ can differ from the true value of $\beta_s^{\rm SM}$:

$$2\beta_s^{J/\psi\phi} = 2\beta_s^{\rm SM} - \phi_s^{\rm NP} \tag{2.24}$$

where from the experimental constraints on the CKM-matrix elements [26], $-2\beta_s^{\rm SM}$ assumes the value

$$2\beta_s^{\rm SM} = 0.0363^{+0.0016}_{-0.0015}. \tag{2.25}$$

With the current experimental sensitivity, the non-SM physics phase would be expected to dominate a measurement of phase. The study of the time evolution of $B_s^0 \to J/\psi\phi$ decays is widely recognized as the best way to probe CP-violation in the interefence between mixing and decay in the B_s^0 sector. The $J/\psi\phi$ final state is common to B_s^0 and $\bar{B}_s^0$ decays, a necessary condition for mixing-induced CP violation to occur. The mixing phase becomes observable through the interference of two amplitudes, the amplitude of direct decay and the amplitude of decay preceded by mixing to a common final-state, Fig. 2.4. What is actually observable is the phase *difference* between decay and mixing, but since the decay is dominated by a single real amplitude, the difference approximates accurately the mixing phase. The fact that the decay is strongly dominated by a single, tree-level, real amplitude is what makes the extraction of the mixing phase from this process theoretically solid. Subleading *penguin* amplitudes are expected to contribuite at the $\mathcal{O}(10^{-3})$ level [27–29], introducing additional phases that in principle complicate the theoretical interpretation of the experimental results. While these effects are completely negligible compared to the expected resolution of the present measurement, they will likely need to be accounted for in the interpretation of future, more precise results.

Because of the spin-composition of the initial- and final-state particles, angular momentum conservation imposes a relative angular momentum (L) between the

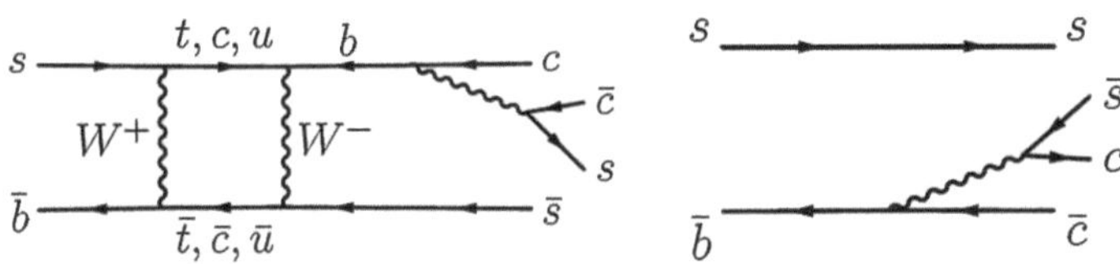

Fig. 2.4 Leading Feynman graph of the $B_s^0 \to J/\psi\phi$ decay with (*left*) and without (*right*) mixing

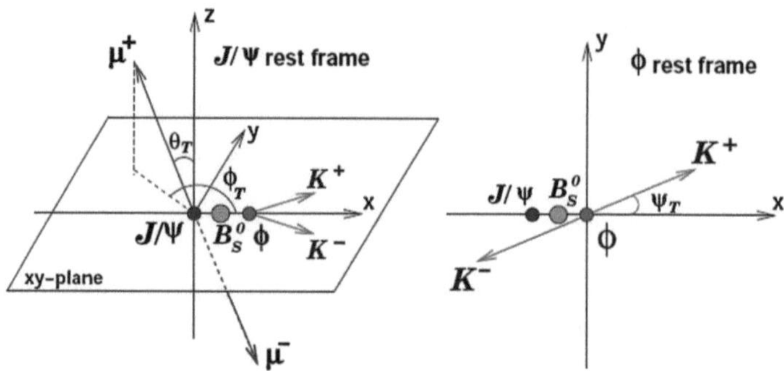

Fig. 2.5 Graphical rappresentation of the relevant angles in the transversity basis

vector (spin-1) final-state particles in order to match the zero net total angular momentum of the initial state. Three independent decay amplitudes determine the transition probability, each corresponding to one of the three possible relative angular momenta, $L = 0, 1, 2$. The transversity basis illustrated in Fig. 2.5, is particularly convenient because when applied to the amplitude, it allows to separate the latter into three terms, each corresponding to a definite CP eigenvalue of the final state, and their interferences. Determining independently the time evolution of decays into CP-even and CP-odd final states enhances the sensitivity to the CP-violating phase, while providing also access to observables, arising from the interference between components with opposite CP parity, that do not vanish even if the evolution of initially produced B_s^0 and $\bar{B}_s^0$ mesons are not separated with flavor-tagging. A candidate-specific determination of the CP parity is not possible, but angular distributions of final-state particles are used to statistically separate CP-even and CP-odd components. Three angles completely define the kinematic distributions of the four final-state particles. In the transversity basis the angles are defined in two different frames. In the following, $\vec{p}(A)_B$ denotes the three momentum of particle A in the rest frame of particle B. The angle Ψ of the K^+ is defined in the ϕ rest frame as the angle between $\vec{p}(K^+)$ and the negative J/ψ direction:

$$\cos \Psi = -\frac{\vec{p}(K^+)_\phi \cdot \vec{p}(J/\psi)_\phi}{|\vec{p}(K^+)_\phi| \cdot |\vec{p}(J/\psi)_\phi|}. \tag{2.26}$$

To calculate the other two angles, we first define a coordinate system through the directions

$$\hat{x} = \frac{\vec{p}(\phi)_{J/\psi}}{|\vec{p}(\phi)_{J/\psi}|}, \quad \hat{y} = \frac{\vec{p}(K^+)_{J/\psi} - [\vec{p}(K^+)_{J/\psi} \cdot \hat{x}]\hat{x}}{|\vec{p}(K^+)_{J/\psi} - [\vec{p}(K^+)_{J/\psi} \cdot \hat{x}]\hat{x}|}, \quad \hat{z} = \hat{x} \times \hat{y}. \tag{2.27}$$

The following angles of the direction of the μ^+ in the J/ψ rest frame are calculated as

$$\cos\Theta = \frac{\vec{p}\,(\mu^+)_{J/\psi}}{|\vec{p}\,(\mu^+)_{J/\psi}|} \cdot \hat{z}, \qquad \Phi = \arctan\left(\frac{\frac{\vec{p}\,(\mu^+)_{J/\psi}}{|\vec{p}\,(\mu^+)_{J/\psi}|}\cdot\hat{y}}{\frac{\vec{p}\,(\mu^+)_{J/\psi}}{|\vec{p}\,(\mu^+)_{J/\psi}|}\cdot\hat{x}}\right) \tag{2.28}$$

where the ambiguity of the angle Φ is lifted by using the signs of the $\vec{p}\,(\mu^+)_{J/\psi}\cdot\hat{x}$ and $\vec{p}\,(\mu^+)_{J/\psi}\cdot\hat{y}$ dot products.

The decay is further described in terms of the polarization states of the vector mesons, either longitudinal (0), or transverse to their directions of motion, and in the latter case, parallel ($\parallel$) or perpendicular ($\perp$) to each other. The corresponding amplitudes, which depend on time t, are A_0, $A_\parallel$ and $A_\perp$, respectively. The transverse linear polarization amplitudes $A_\parallel$ and $A_\perp$ correspond to CP-even and CP-odd final states at decay time $t = 0$, respectively. The longitudinal polarization amplitude A_0 corresponds to a CP-even final state. The angular distribution of $B_s^0 \rightarrow J/\phi(\rightarrow \mu^+\mu^-)\phi(\rightarrow K^+K^-)$ decays reads,

$$\frac{1}{\overset{(-)}{\Gamma}}\frac{\mathrm{d}^3\,\overset{(-)}{\Gamma}\,(\overset{(-)}{B}{}_s^0 \rightarrow J/\psi\phi)}{\mathrm{d}\cos\Theta\,\mathrm{d}\Phi\,\mathrm{d}\cos\Psi} = \frac{\sum_{i=1}^{6}\left[\overset{(-)}{K}_i\,f_i(\cos\Theta,\Phi,\cos\Psi)\right]}{|\overset{(-)}{A}_0|^2 + |\overset{(-)}{A}_\parallel|^2 + |\overset{(-)}{A}_\perp|^2}, \tag{2.29}$$

with the K_i and $f_i(\cos\Theta,\Phi,\cos\Psi)$ terms detailed in Table 2.2. Figure 2.6 shows some examples of transversity-angles distributions in $B_s^0 \rightarrow J/\psi\phi$ decays for three sets of polarization amplitudes, illustrating how the distribution of these observables depend on the underlying physics parameters.

Effects from $B_s^0 - \bar{B}_s^0$ oscillations are introduced along with the decay transition to the final state. The time evolution is independent of the angular distributions, and it is encoded through a time-dependence of the polarization amplitudes, i.e., the K_i terms of Eq. (2.29):

$$K_i \rightarrow K_i(t),$$

Table 2.2 Angular functions in terms of transversity angles and corresponding transversity amplitudes for the $B_s^0 \rightarrow J/\psi\phi$ decay entering Eq. (2.29)

i	$\overset{(-)}{K}_i$	$f_i(\cos\Theta,\Phi,\cos\Psi)$
1	$\|\overset{(-)}{A}_0\|^2$	$\frac{9}{32\pi}2\cos^2\Psi(1-\sin^2\Theta\cos^2\Phi)$
2	$\|\overset{(-)}{A}_\parallel\|^2$	$\frac{9}{32\pi}\sin^2\Psi(1-\sin^2\Theta\sin^2\Phi)$
3	$\|\overset{(-)}{A}_\perp\|^2$	$\frac{9}{32\pi}\sin^2\Psi\sin^2\Theta$
4	$\Im\mathrm{m}(\overset{(-)}{A}_\perp\,\overset{(-)}{A}_\parallel^\star)$	$-\frac{9}{32\pi}\sin^2\Psi\sin 2\Theta\sin\Phi$
5	$\Re\mathrm{e}(\overset{(-)}{A}_\parallel\,\overset{(-)}{A}_0^\star)$	$\frac{9}{32\pi}\frac{\sqrt{2}}{2}\sin 2\Psi\sin^2\Theta\sin 2\Phi$
6	$\Im\mathrm{m}(\overset{(-)}{A}_\perp\,\overset{(-)}{A}_0^\star)$	$\frac{9}{32\pi}\frac{\sqrt{2}}{2}\sin 2\Psi\sin 2\Theta\cos\Phi$

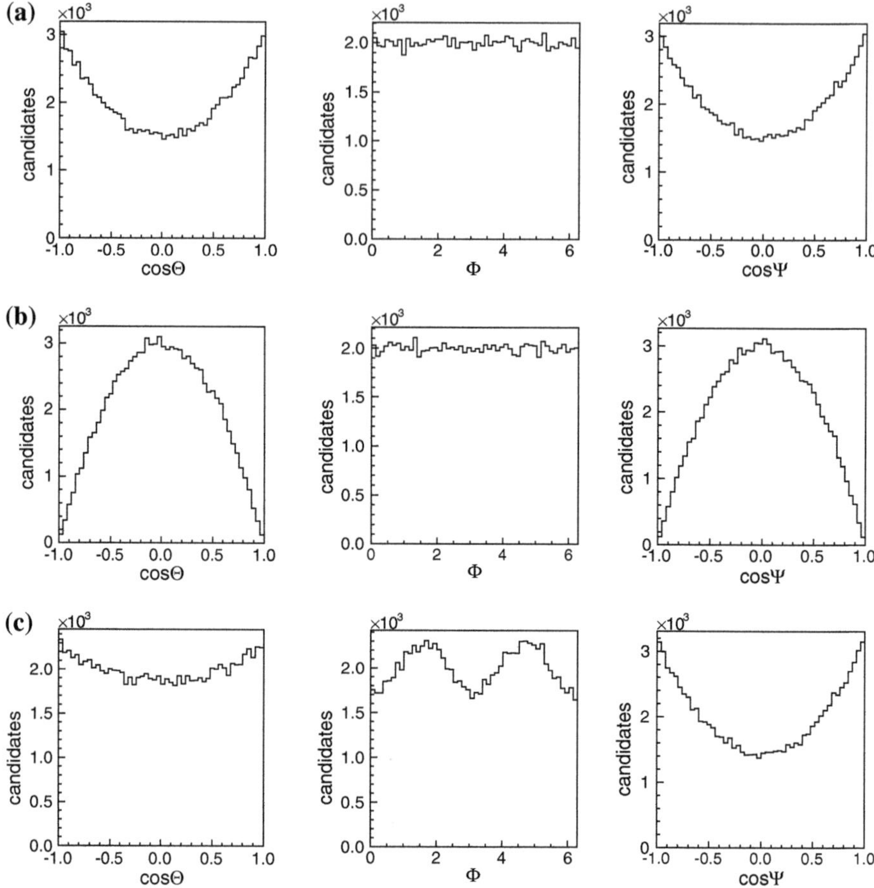

Fig. 2.6 Example of transversity angles distributions. In **a** the CP-odd amplitude $A_\perp$ is set zero, while both $|A_0|^2$ and $|A_\parallel|^2$ are set to 50 %. In **b** the CP-even amplitudes A_0 and $A_\perp$ are set to zero. In **c** the amplitudes and strong phases are set to the value measured in Ref. [30]: $|A_0|^2 = 0.524$, $|A_\parallel|^2 = 0.231, \delta_\perp = 2.95$, and we take $\delta_\parallel = \pi$. Note that the angle Φ has a non-uniform distribution if there is an interference between the CP-even and CP-odd amplitudes

where t is the decay-time. By using Eqs. (2.14) and (2.16) and considering the CP parity of each transversity amplitude, one can derive the time development for each $K_i(t)$ term. The decay rate of an initially produced $\overset{(-)}{B}{}^0_s$ meson is written as a function of the decay time and transversity angles as in Ref. [31]. In Table 2.3, we list the $\overset{(-)}{K}{}_i(t)$ terms, where the polarization amplitudes at $t = 0$ for an initially produced $\overset{(-)}{B}{}^0_s$ meson are defined as

Table 2.3 General expressions of $K_i(t)$ ($\bar{K}_i(t)$) terms for $B_s^0 \to J/\psi\phi$ decays, where both direct and interference CP violation are allowed

i	$\overset{(-)}{K}_i$
1	$\frac{1}{2}e^{-\Gamma t}\Big[(|A_0|^2+|\bar{A}_0|^2)\cosh(\Delta\Gamma_s t/2)\overset{(-)}{+}(|A_0|^2-|\bar{A}_0|^2)\cos\Delta m_s t$ $\quad-2\Re e(A_0^\star\bar{A}_0)\big(\cos2\phi_M^s\sinh(\Delta\Gamma_s t/2)\overset{(+)}{-}\sin2\phi_M^s\sin\Delta m_s t\big)$ $\quad-2\Im m(A_0^\star\bar{A}_0)\big(\overset{(-)}{+}\cos2\phi_M^s\sin\Delta m_s t+\sin2\phi_M^s\sinh(\Delta\Gamma_s t/2)\big)\Big]$
2	$\frac{1}{2}e^{-\Gamma t}\Big[(|A_\parallel|^2+|\bar{A}_\parallel|^2)\cosh(\Delta\Gamma_s t/2)\overset{(-)}{+}(|A_\parallel|^2-|\bar{A}_\parallel|^2)\cos\Delta m_s t$ $\quad-2\Re e(A_\parallel^\star\bar{A}_\parallel)\big(\cos2\phi_M^s\sinh(\Delta\Gamma_s t/2)\overset{(+)}{-}\sin2\phi_M^s\sin\Delta m_s t\big)$ $\quad-2\Im m(A_\parallel^\star\bar{A}_\parallel)\big(\overset{(-)}{+}\cos2\phi_M^s\sin\Delta m_s t+\sin2\phi_M^s\sinh(\Delta\Gamma_s t/2)\big)\Big]$
3	$\frac{1}{2}e^{-\Gamma t}\Big[(|A_\perp|^2+|\bar{A}_\perp|^2)\cosh(\Delta\Gamma_s t/2)\overset{(-)}{+}(|A_\perp|^2-|\bar{A}_\perp|^2)\cos\Delta m_s t$ $\quad+2\Re e(A_\perp^\star\bar{A}_\perp)\big(\cos2\phi_M^s\sinh(\Delta\Gamma_s t/2)\overset{(+)}{-}\sin2\phi_M^s\sin\Delta m_s t\big)$ $\quad+2\Im m(A_\perp^\star\bar{A}_\perp)\big(\overset{(-)}{+}\cos2\phi_M^s\sin\Delta m_s t+\sin2\phi_M^s\sinh(\Delta\Gamma_s t/2)\big)\Big]$
4	$\frac{1}{2}e^{-\Gamma t}\Big[\Im m(A_\perp A_\parallel^\star-\bar{A}_\perp\bar{A}_\parallel^\star)\cosh(\Delta\Gamma_s t/2)\overset{(-)}{+}\Im m(A_\perp A_\parallel^\star+\bar{A}_\perp\bar{A}_\parallel^\star)\cos\Delta m_s t$ $\quad+\Im m(A_\perp\bar{A}_\parallel^\star-\bar{A}_\perp A_\parallel^\star)\big(-\sinh(\Delta\Gamma_s t/2)\cos2\phi_M^s\overset{(-)}{+}\sin\Delta m_s t\sin2\phi_M^s\big)$ $\quad+\Re e(A_\perp\bar{A}_\parallel^\star+\bar{A}_\perp A_\parallel^\star)\big(-\sinh(\Delta\Gamma_s t/2)\sin2\phi_M^s\overset{(+)}{-}\sin\Delta m_s t\cos2\phi_M^s\big)\Big]$
5	$\frac{1}{2}e^{-\Gamma t}\Big[\Re e(A_\parallel A_0^\star+\bar{A}_\parallel\bar{A}_0^\star)\cosh(\Delta\Gamma_s t/2)\overset{(-)}{+}\Re e(A_\parallel A_0^\star-\bar{A}_\parallel\bar{A}_0^\star)\cos\Delta m_s t$ $\quad+\Re e(A_\parallel\bar{A}_0^\star+\bar{A}_\parallel A_0^\star)\big(-\sinh(\Delta\Gamma_s t/2)\cos2\phi_M^s\overset{(-)}{+}\sin\Delta m_s t\sin2\phi_M^s\big)$ $\quad+\Im m(A_\parallel\bar{A}_0^\star-\bar{A}_\parallel A_0^\star)\big(\sinh(\Delta\Gamma_s t/2)\sin2\phi_M^s\overset{(-)}{+}\sin\Delta m_s t\cos2\phi_M^s\big)\Big]$
6	$\frac{1}{2}e^{-\Gamma t}\Big[\Im m(A_\perp A_0^\star-\bar{A}_\perp\bar{A}_0^\star)\cosh(\Delta\Gamma_s t/2)\overset{(-)}{+}\Im m(A_\perp A_0^\star+\bar{A}_\perp\bar{A}_0^\star)\cos\Delta m_s t$ $\quad+\Im m(A_\perp\bar{A}_0^\star-\bar{A}_\perp A_0^\star)\big(-\sinh(\Delta\Gamma_s t/2)\cos2\phi_M^s\overset{(-)}{+}\sin\Delta m_s t\sin2\phi_M^s\big)$ $\quad+\Re e(A_\perp\bar{A}_0^\star+\bar{A}_\perp A_0^\star)\big(-\sinh(\Delta\Gamma_s t/2)\sin2\phi_M^s\overset{(+)}{-}\sin\Delta m_s t\cos2\phi_M^s\big)\Big]$

$$\overset{(-)}{A}_0(t=0) = \overset{(-)}{A}_0 = \langle J/\psi\phi, 0|\,\overset{(-)}{B}{}_s^0\rangle,$$

$$\overset{(-)}{A}_\parallel(t=0) = \overset{(-)}{A}_\parallel = \langle J/\psi\phi, \parallel\,|\,\overset{(-)}{B}{}_s^0\rangle, \tag{2.30}$$

$$\overset{(-)}{A}_\perp(t=0) = \overset{(-)}{A}_\perp = \langle J/\psi\phi, \perp\,|\,\overset{(-)}{B}{}_s^0\rangle,$$

The expressions in Table 2.3 refer to the most general case for the time evolution of a B_s^0-decay into a vector–vector and self-conjugate final state. By constructing the quantity

$$\frac{K_i(t) - \bar{K}_i(t)}{K_i(t) + \bar{K}_i(t)} \tag{2.31}$$

with $i = (1, 2, 3, 5)$, one can obtain the time-dependent CP asymmetry

$$\mathcal{A}_{CP}(t) = \frac{\Gamma(B_s^0 \to \bar{f}) - \Gamma(\bar{B}_s^0 \to f)}{\Gamma(B_s^0 \to \bar{f}) + \Gamma(\bar{B}_s^0 \to f)}$$
$$\simeq \frac{\mathcal{C}_f \cos(\Delta m_s t) - \mathcal{S}_f \sin(\Delta m_s t)}{\cosh(\Delta \Gamma_s t/2) + \mathcal{S}'_f \sinh(\Delta \Gamma_s t/2)}. \tag{2.32}$$

where

$$\mathcal{C}_f = \frac{1 - |\lambda_f|^2}{1 + |\lambda_f|^2}, \quad \mathcal{S}_f = \frac{2\Im\mathfrak{m}(\lambda_f)}{1 + |\lambda_f|^2}, \quad \mathcal{S}'_f = \frac{2\Re\mathfrak{e}(\lambda_f)}{1 + |\lambda_f|^2}, \tag{2.33}$$

This includes both *direct* and *interference CP* violation.

In particular, the polarization amplitudes are written as follows:

$$\overset{(-)}{A_0} = |\overset{(-)}{A_0}|e^{\overset{(-)}{+}i\phi_{A_0}}e^{i\delta_0},$$

$$\overset{(-)}{A_\parallel} = |\overset{(-)}{A_\parallel}|e^{\overset{(-)}{+}i\phi_{A_\parallel}}e^{i\delta_\parallel},$$

$$\overset{(-)}{A_\perp} = |\overset{(-)}{A_\perp}|e^{\overset{(-)}{+}i\phi_{A_\perp}}e^{i\delta_\perp}, \tag{2.34}$$

with ϕ_{A_i} and δ_i being the weak and strong phases of the amplitudes, respectively. Table 2.3 emphasizes the two distinct sources of *CP* violation. The terms proportional to $\Re\mathfrak{e}(\overset{(-)}{A_i}\,\overset{(-)}{A_j}{}^\star)$ and to $\Im\mathfrak{m}(\overset{(-)}{A_i}\,\overset{(-)}{A_j}{}^\star)$ in the $\overset{(-)}{K}_i(t)$ terms encode the dependence on both ϕ_{A_i} and δ_i. The time-evolution functions $K_i(t)$ are reported in Table 2.4 in a compact form that emphasizes the equality of the time dependence for final states with the same *CP*-parity. We have defined

Table 2.4 Expressions of $K_i(t)$ terms of the $B_s^0 \to J/\psi\phi$ decay rate, where only interference *CP* violation is allowed

i	$\overset{(-)}{K}_i(t)$	CP parity
1	$\|\overset{(-)}{A}_0\|^2 \overset{(-)}{\mathcal{O}}{}^+(t)$	Even
2	$\|\overset{(-)}{A}_\parallel\|^2 \overset{(-)}{\mathcal{O}}{}^+(t)$	Even
3	$\|\overset{(-)}{A}_\perp\|^2 \overset{(-)}{\mathcal{O}}{}^-(t)$	Odd
4	$\|\overset{(-)}{A}_\parallel\|\|\overset{(-)}{A}_\perp\| \overset{(-)}{\mathcal{E}}_{\Im\mathrm{m}}(t, \delta_\perp - \delta_\parallel)$	Mix
5	$\|\overset{(-)}{A}_\parallel\|\|\overset{(-)}{A}_0\|\cos\delta_\parallel \overset{(-)}{\mathcal{O}}{}^+(t)$	Even
6	$\|\overset{(-)}{A}_\perp\|\|\overset{(-)}{A}_0\| \overset{(-)}{\mathcal{E}}_{\Im\mathrm{m}}(t, \delta_\perp)$	Mix

The third column reports the *CP* parity of each term. The formulae of $\mathcal{O}^\pm$ and $\mathcal{E}_{\Im\mathrm{m}}$ are given by Eqs. (2.35) and (2.36), while the phases $\delta_\parallel$ and $\delta_\perp$ are defined by Eq. (2.37)

$$
\overset{(-)}{\mathcal{O}}{}^{\pm}(t) = e^{-\Gamma t}\left(\cosh \frac{\Delta\Gamma_s t}{2} \mp \cos 2\beta_s^{J/\psi\phi} \sinh \frac{\Delta\Gamma_s t}{2} \overset{(\mp)}{\pm} \sin 2\beta_s^{J/\psi\phi} \sin \Delta m_s t \right),
$$

$$
\tag{2.35}
$$

$$
\overset{(-)}{\mathcal{E}}{}_{\Im m}(t,\alpha) = e^{-\Gamma t}\left(\overset{(-)}{+} \sin\alpha \cos \Delta m_s t \overset{(+)}{-} \cos\alpha \cos 2\beta_s^{J/\psi\phi} \sin \Delta m_s t \right.
$$

$$
\left. - \cos\alpha \sin 2\beta_s^{J/\psi\phi} \sinh \frac{\Delta\Gamma_s t}{2} \right).
\tag{2.36}
$$

The phase α in the above equations represents the *CP*-conserving phase associated with the polarization amplitudes. Since only phase differences matter, a customary convention is to choose A_0 real and define the strong phases of the transverse amplitudes as

$$
\delta_{\parallel} = \arg\left(\frac{A_{\parallel}}{A_0}\right) = \arg\left(\frac{\bar{A}_{\parallel}}{\bar{A}_0}\right),
\tag{2.37}
$$

$$
\delta_{\perp} = \arg\left(\frac{A_{\perp}}{A_0}\right) = \arg\left(\frac{\bar{A}_{\perp}}{\bar{A}_0}\right).
$$

2.4.1 Time Evolution of the $(S+P)$-wave System

Thus far we only considered K^+K^- pairs originated from the decay of $\phi(1020)$ mesons. However, the K^+K^- final state could include a mixture of multiple resonances, their interference, and also contribution from non-resonant production. Neglecting the contamination of $s\bar{s}$-quark states of zero spin and mass close to the $\phi(1020)$ pole, such as the $f_0(980)$ [7], may induce a bias in the estimation of the *CP*-odd fraction of the signal and alter the measurement of $\beta_s^{J/\psi\phi}$ [32].

We now focus on the decay $B_s^0 \to J/\psi X(\to K^+K^-)$. The $f_0(980)$ is a spin-0 meson that may contribute and its contribution is called often *S-wave*, while the $\phi(1020)$ is a spin-1 meson and its contribution is denoted as *P-wave*. A fraction of $B_s^0 \to J/\psi K^+K^-$ decays may be present with a non-resonant K^+K^- pair that has a relative angular momentum $L=1$ with respect the J/ψ state. Such contributions are expected to be significantly smaller than the resonant fraction, because their amplitude involves the production of an extra $u\bar{u}$-quark pair with respect to the amplitude for the resonances' production. Hence, the non-resonant component is not further considered. The partial-wave classification of the various contributions to the K^+K^- spectrum, which is based on the spin of the resonance, should not be confused with the partial-wave basis of the polarization amplitudes, which is based on the value of the relative angular momentum between two resonances. In what follows, the partial-waves nomenclature is used only for referring to the K^+K^-

resonances spectrum, while for polarization amplitudes only the transversity basis is used.

The differential decay rates considered so far—e.g., Eq. (2.29)—are parametrized as functions of transversity angles and decay time. However, they also depend on the invariant K^+K^- mass m, resulting in amplitudes that are functions of m as well [33, 34]. To account for the total $S + P$ contribution in the decay rate, the P-wave amplitude and the S-wave amplitude are summed; then the total decay rate is decomposed as follows [33, 34]:

$$
\frac{\mathrm{d}^5 \overset{(-)}{\Gamma}(\overset{(-)}{B}{}^0_s \to J/\psi K^+ K^-)}{\mathrm{d}m\,\mathrm{d}t\,\mathrm{d}\cos\Theta\,\mathrm{d}\Phi\,\mathrm{d}\cos\Psi} = |\overset{(-)}{P}{}_{\text{wave}} + \overset{(-)}{S}{}_{\text{wave}}|^2
$$

$$
= |\overset{(-)}{P}{}_{\text{wave}}|^2 + |\overset{(-)}{S}{}_{\text{wave}}|^2 + 2\,\mathfrak{Re}\left(\overset{(-)}{P}{}_{\text{wave}}\,\overset{(-)}{S}{}^{\star}_{\text{wave}}\right),
$$

$$(2.38)$$

Usually, in analyses of $B^0_s \to J/\psi\phi$ decays, the dependence on m is integrated out and the measurement of the polarization amplitudes is obtained from angular distributions only, obtaining

$$
\frac{1}{\overset{(-)}{\Gamma}}\frac{\mathrm{d}^4\overset{(-)}{\Gamma}(\overset{(-)}{B}{}^0_s \to J/\psi K^+K^-)}{\mathrm{d}t\,\mathrm{d}\cos\Theta\,\mathrm{d}\Phi\,\mathrm{d}\cos\Psi} = \frac{\sum_{i=1}^{10}\left[\overset{(-)}{K}{}_i(t)\,f_i(\cos\Theta,\Phi,\cos\Psi)\right]}{|\overset{(-)}{A}{}_0|^2 + |\overset{(-)}{A}{}_\parallel|^2 + |\overset{(-)}{A}{}_\perp|^2 + |\overset{(-)}{A}{}_S|^2},
$$

$$(2.39)$$

Table 2.5 Expressions of $K_i(t)$ ($\bar{K}_i(t)$) and $f_i(\cos\Theta, \Phi, \cos\Psi)$ terms of the $B^0_s \to J/\psi K^+K^-$ decay rate, where the dependence on the mass m is integrated out

i	$\overset{(-)}{K}{}_i(t)$	$f_i(\cos\Theta, \Phi, \cos\Psi)$	CP parity
1	$\|\overset{(-)}{A}{}_0\|^2\,\overset{(-)}{\mathcal{O}}{}^+(t)$	$\frac{9}{32\pi}2\cos^2\Psi(1 - \sin^2\Theta\cos^2\Phi)$	Even
2	$\|\overset{(-)}{A}{}_\parallel\|^2\,\overset{(-)}{\mathcal{O}}{}^+(t)$	$\frac{9}{32\pi}\sin^2\Psi(1 - \sin^2\Theta\sin^2\Phi)$	Even
3	$\|\overset{(-)}{A}{}_\perp\|^2\,\overset{(-)}{\mathcal{O}}{}^-(t)$	$\frac{9}{32\pi}\sin^2\Psi\sin^2\Theta$	Odd
4	$\|\overset{(-)}{A}{}_\parallel\|\|\overset{(-)}{A}{}_\perp\|\,\overset{(-)}{\mathcal{E}}{}_{\mathfrak{Im}}(t,\delta_\perp-\delta_\parallel)$	$-\frac{9}{32\pi}\sin^2\Psi\sin 2\Theta\sin\Phi$	Mix
5	$\|\overset{(-)}{A}{}_\parallel\|\|\overset{(-)}{A}{}_0\|\cos\delta_\parallel\,\overset{(-)}{\mathcal{O}}{}^+(t)$	$\frac{9}{32\pi}\frac{\sqrt{2}}{2}\sin 2\Psi\sin^2\Theta\sin 2\Phi$	Even
6	$\|\overset{(-)}{A}{}_\perp\|\|\overset{(-)}{A}{}_0\|\,\overset{(-)}{\mathcal{E}}{}_{\mathfrak{Im}}(t,\delta_\perp)$	$\frac{9}{32\pi}\frac{\sqrt{2}}{2}\sin 2\Psi\sin 2\Theta\cos\Phi$	Mix
7	$\|\overset{(-)}{A}{}_S\|^2\,\overset{(-)}{\mathcal{O}}{}^-(t)$	$\frac{3}{32\pi}2(1 - \sin^2\Theta\cos^2\Phi)$	Odd
8	$I_m\|\overset{(-)}{A}{}_\parallel\|\|\overset{(-)}{A}{}_S\|\,\overset{(-)}{\mathcal{E}}{}_{\mathfrak{Re}}(t,\delta_\parallel-\delta_S)$	$\frac{3}{32\pi}2\cos\Psi(1 - \sin^2\Theta\cos^2\Phi)$	Mix
9	$I_m\|\overset{(-)}{A}{}_\perp\|\|\overset{(-)}{A}{}_S\|\sin(\delta_\perp-\delta_S)\,\overset{(-)}{\mathcal{O}}{}^-(t)$	$\frac{3}{32\pi}\frac{1}{\sqrt{2}}2\sin\Psi\sin^2\Theta\sin 2\Phi$	Odd
10	$I_m\|\overset{(-)}{A}{}_0\|\|\overset{(-)}{A}{}_S\|\,\overset{(-)}{\mathcal{E}}{}_{\mathfrak{Re}}(t,-\delta_S)$	$\frac{3}{32\pi}\frac{1}{\sqrt{2}}2\sin\Psi\sin 2\Theta\cos\Phi$	Mix

The last column reports the CP parity of each term. The formulae of $\mathcal{O}^\pm$ and $\mathcal{E}_{\mathfrak{Im}}$ are given by Eqs. (2.35) and (2.40). The coefficient I_m is the integral of the Breit-Wigner resonance mass distribution times the S-*wave* component line shape distribution

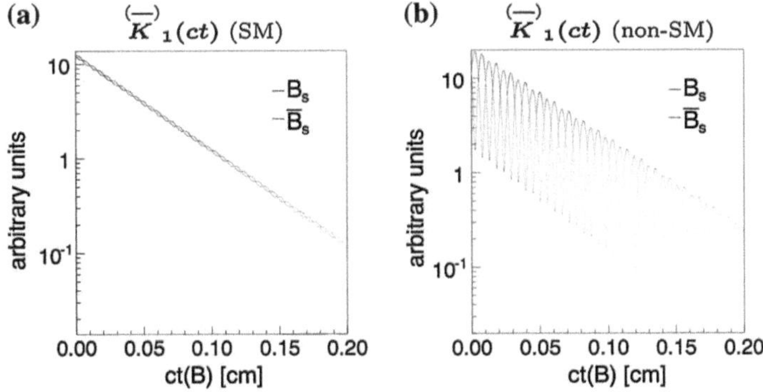

Fig. 2.7 Evolution of the amplitude K_1 for simulated $B_s^0 \to J/\psi\, K^+ K^-$ decays according to Eq. (2.39), as a function of ct. *The blue line* refers to initially produced B_s^0 mesons and *red line* is for $\bar{B}_s^0$. In **a** the values $\beta_s = 0.02$ and $\Delta\Gamma_s = 0.09\,\mathrm{ps}^{-1}$ (SM point) are used, while in **b** $\beta_s = 0.5$ and $\Delta\Gamma_s = 0.09\cos(2\beta_s^{J/\psi\phi}) = 0.049\,\mathrm{ps}^{-1}$. In both cases, $\Delta m_s = 17.77\,\mathrm{ps}^{-1}$

with the expressions of $\overset{(-)}{K}_i(t)$ and $f_i(\cos\Theta, \Phi, \cos\Psi)$ given in Table 2.5, where the following shorthand in the time-evolution of the interference term between P- and S-wave with mixed CP-parity is introduced:

$$\overset{(-)}{\mathcal{E}}_{\mathfrak{Re}}(t, \alpha) = e^{-\Gamma t}\left(\overset{(-)}{+} \cos\alpha \cos\Delta m_s t \overset{(+)}{-} \sin\alpha \cos 2\beta_s^{J/\psi\phi} \sin\Delta m_s t \right. \tag{2.40}$$

$$\left. - \sin\alpha \sin 2\beta_s^{J/\psi\phi} \sinh\frac{\Delta\Gamma_s t}{2} \right),$$

where α identifies various combinations of strong phases, as indicated in Table 2.5. Figure 2.7 sketches the evolution of the amplitude K_1 ($|A_0(t)|^2$) of the $B_s^0 \to J/\psi\, K^+ K^-$ decay rate in Eq. (2.39), as a function of the decay-length, separately for the B_s^0 and the $\bar{B}_s^0$ mesons, assuming the values of $\beta_s^{J/\psi\phi} = 0.02$ and $\Delta\Gamma_s = 0.90\,\mathrm{ps}^{-1}$ in Fig. 2.7a, and $\beta_s^{J/\psi\phi} = 0.5$ and $\Delta\Gamma_s = 0.09\cos(2\beta_s^{J/\psi\phi}) = 0.49\,\mathrm{ps}^{-1}$ in Fig. 2.7b. The value of the oscillation frequency is fixed to the known value $\Delta m_s = 17.77\,\mathrm{ps}^{-1}$ [35], and polarization amplitudes and strong phases are as measured in Ref. [30]. The time-evolution of the decay-amplitude changes significantly for different values of $\beta_s^{J/\psi\phi}$ and $\Delta\Gamma_s$. Specifically, the squared magnitudes of the polarization amplitudes depend on the terms $\cos 2\beta_s^{J/\psi\phi} \sinh(\Delta\Gamma_s t/2)$ and $\sin 2\beta_s^{J/\psi\phi} \sin(\Delta m_s t)$; the former provides sensitivity to $\beta_s^{J/\psi\phi}$ even without distinction of the flavor of the B_s^0-meson at production, if $\Delta\Gamma_s$ differs from zero. This last term also explains the different size of the oscillations amplitude between the two cases.

2.4.2 Likelihood Symmetries

The decay rate in Eq. (2.39) features some symmetries, i.e., transformations of some of the observables of interest that leave the equations invariant. We first consider the simpler case, where only the P-wave is present. Assuming that B_s^0 and the $\bar{B}_s^0$ mesons are not distinguished at production, and that they are produced in equal amount (*untagged* sample), then the $B_s^0 \to J/\psi K^+ K^-$ and the $\bar{B}_s^0 \to J/\psi K^+ K^-$ decay rates are summed. Each oscillation term proportional to $\sin \Delta m_s t$ or $\cos \Delta m_s t$ is canceled out because they appear with opposite sign in the $K_i(t)$ and $\bar{K}_i(t)$ terms, but the rate is still sensitive to $\beta_s^{J/\psi\phi}$ if $\Delta\Gamma_s \neq 0$. The untagged decay rate is invariant under the parameter transformation [34]

$$\begin{cases} \beta_s^{J/\psi\phi} \to \pi/2 - \beta_s^{J/\psi\phi} \\ \Delta\Gamma_s \to -\Delta\Gamma_s \\ \delta_\parallel \to 2\pi - \delta_\parallel \\ \delta_\perp \to \pi - \delta_\perp \end{cases} \tag{2.41}$$

together with the reflection of this transformation with respect to $\beta_s^{J/\psi\phi} = 0$

$$\begin{cases} \beta_s^{J/\psi\phi} \to -\beta_s^{J/\psi\phi} \\ \Delta\Gamma_s \to \Delta\Gamma_s \\ \delta_\parallel \to \delta_\parallel \\ \delta_\perp \to \delta_\perp \end{cases} \quad \text{and} \quad \begin{cases} \pi/2 - \beta_s^{J/\psi\phi} \to -\pi/2 + \beta_s^{J/\psi\phi} \\ -\Delta\Gamma_s \to -\Delta\Gamma_s \\ 2\pi - \delta_\parallel \to 2\pi - \delta_\parallel \\ \pi - \delta_\perp \to \pi - \delta_\perp. \end{cases} \tag{2.42}$$

A four-fold ambiguity is present in the decay rate, with the four equivalent solutions sketched in the $(\beta_s^{J/\psi\phi}, \Delta\Gamma_s)$ plane in Fig. 2.8.

Fig. 2.8 Example in the $(\beta_s^{J/\psi\phi}, \Delta\Gamma_s)$ plane of the equivalent values of $\beta_s^{J/\psi\phi}$, $\Delta\Gamma_s$, and strong phases, that leave the decay rate invariant. The confidence regions at 68 % C.L. (*blue*) and 95 % C.L. (*red*) for the analysis of a pseudo-experiment with (*bold lines*) and without (*light lines*) using the tagging information

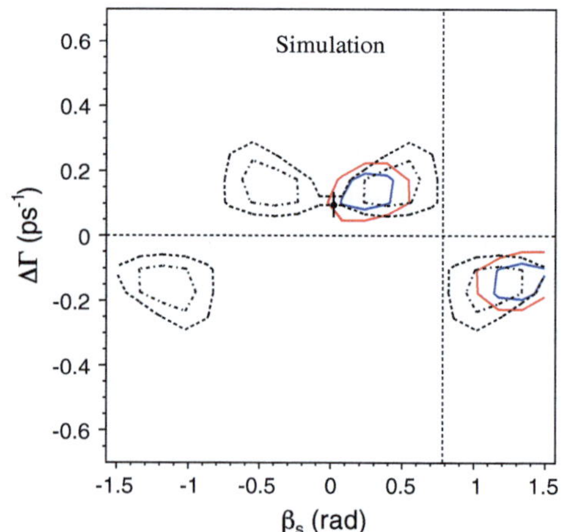

When differences of decay-rates of initially-produced B_s^0 and $\bar{B}_s^0$ meson are included through flavor-tagging, the transformations in Eq. (2.42) are no longer symmetries of the decay rate, and only the transformation of Eq. (2.41) leaves the decay rate invariant, resulting in a two-fold ambiguity (with the cancellation of the solutions for $\beta_s^{J/\psi\phi} < 0$ in Fig. 2.8). Tagging allows indeed to access the following terms of the decay rate:

$$\sin 2\beta_s^{J/\psi\phi} \sin(\Delta m_s t) \quad \text{and} \quad \cos 2\beta_s^{J/\psi\phi} \sin(\Delta m_s t), \tag{2.43}$$

that are not present in the untagged rate. The main effect of the tagging in this analysis is to break the $\beta_s^{J/\psi\phi} \to -\beta_s^{J/\psi\phi}$ symmetry (Sect. 2.4.2), removing half of the allowed region in the $(\beta_s^{J/\psi\phi}, \Delta\Gamma_s)$ space. However, the tagging power is not large enough to substantially reduce the uncertainties on the $\beta_s^{J/\psi\phi}$ estimation of the remaining solutions, and each of the four untagged solutions has comparable uncertainties to those on the tagged solutions, as shown in Fig. 2.8, where we compare the results in the $(\beta_s^{J/\psi\phi}, \Delta\Gamma_s)$ plane for tagged and untagged analysis of one simulated samples of $B_s^0 \to J/\psi\phi$ decays. If the tagging power was greater, we would expect the sensitivity to $\beta_s^{J/\psi\phi}$ to be substantially better in the tagged case, and the uncertainties on $\beta_s^{J/\psi\phi}$ to be smaller.

The contribution of the S-wave state adds a symmetry transformation in the integrated decay rate

$$\begin{cases} \beta_s^{J/\psi\phi} \to \pi/2 - \beta_s^{J/\psi\phi} \\ \Delta\Gamma_s \to -\Delta\Gamma_s \\ \delta_\parallel \to 2\pi - \delta_\parallel \\ \delta_\perp \to \pi - \delta_\perp \\ \delta_S \to \pi - \delta_S. \end{cases} \tag{2.44}$$

The invariance under tranformation Eq. (2.44) requires the symmetry of the K^+K^- resonances mass shape around the $\phi(1020)$ pole. In the case of asymmetric shapes, the transformation Eq. (2.44) leads to an approximate symmetry, which is more approximate as larger becomes the S-wave fraction in the sample. An S-wave fraction as expected in CDF data ($\approx 1\,\%$), leaves the decay rate nearly symmetric under the transformation of Eq. (2.44).

These mathematical features of the decay rates lead to difficulties in the likelihood minimization due to the presence of multiple, equivalent minima (Chap. 6). Proper statistical treatment of the resulting features is an important part of this analysis. The statistical reliability of results that are very sensitive to non-SM physics has to be accurately ensured.

Table 2.6 Experimental status of B mixing observables and corresponding SM predictions in 2010

Observable	Measurement	Source	SM prediction	References
B_s^0 system				
Δm_s (ps^{-1})	17.78 ± 0.12	HFAG 2010 [25]	17.3 ± 2.6	[26, 36–39]
$\Delta \Gamma_s$ (ps^{-1})	0.075 ± 0.035	HFAG 2010 [25]	0.096 ± 0.039	[26, 36–39]
ϕ_s (rad)	$-0.75^{+0.32}_{-0.21}$	HFAG 2010 [25]	-0.036 ± 0.002	[36–41]
a_{sl}^s (10^{-4})	$-17 \pm 91 \, ^{+14}_{-15}$	D0 (no A_{SL}^b) [42]	$0.29^{+0.09}_{-0.08}$	[36–41]
Admixture of B^0 and B_s mesons				
A_{SL}^b (10^{-4})	$-78.7 \pm 17.2 \pm 9.3$	D0 [43]	-2.0 ± 0.3	[26, 36–39]

The inclusive same-sign dimuon asymmetry A_{SL}^b is defined in Ref. [43]

2.5 Experimental Status

The status of measurements and SM predictions for the mixing observables prior this measurement are summarized in Table 2.6.

The world's average value of the B_s^0 mass difference Δm_s in Table 2.6 is based on measurements performed at CDF [35] and LHCb [44]. These are all consistent with the SM predictions within sizable theoretical uncertainties. Improving the precision of the SM prediction is desirable to further constrain non-SM physics in M_{12}^s, and requires improving the accuracy of lattice QCD evaluations of the decay constant and *bag parameter* (see Ref. [26] and references therein).

The first measurements of the *CP*-violating phase in flavor-tagged $B_s^0 \rightarrow J/\psi \phi$ decays was finalized in 2008 by the CDF experiment [45]. It showed a mild, 1.5σ discrepancy from the SM. Intriguingly the D0 experiment found a similar, and consistent effect [46] a few months later. The combination yielded a 2.2σ deviation from the SM [47]. This attracted some interest, further enhanced by the like-sign dimuon asymmetry results from the D0 collaboration [43]. Such asymmetry, A_{sl}^b, receives contributions from the flavor-specific asymmetries in B^0 and B_s^0 semileptonic decays, a_{sl}^d and a_{sl}^s, respectively. The large value of A_{sl}^b observed by D0, combined with precise determinations of a_{sl}^d from the B factories, suggested an anomalous value of a_{sl}^s and thus an anomalous value of the phase ϕ_{12}^s, which can be accurately tested using $B_s^0 \rightarrow J/\psi \phi$ decays. In 2010, both the CDF and D0 collaborations updated their measurements of $B_s^0 \rightarrow J/\psi \phi$ time-evolution using events sample based on 5.2 fb^{-1} and 8 fb^{-1} of integrated luminosity [30, 48]. The results, although consistent with the previous ones, showed an improved agreement with the SM. Also LHCb began to contribute, with a measurement based 340 pb^{-1} of data [44], which already had competitive precision.

2.6 Experimental Aspects

The $B_s^0 \to J/\psi(\to \mu^+\mu^-)\phi(\to K^+K^-)$ decay is considered one of the handful of *golden channels* in flavor physics. In addition to allowing access to a key observable, sensitive to a broad class of non-SM physics models and reliably predicted, it offers several experimental advantages. The combined branching fraction is at the 10^{-5} level, which makes the collection of significant samples possible in hadron collisions. All final-state particles are charged, thus easier to reconstruct in hadron collisions. In particular, the two muons in the final states originating from the narrow J/ψ resonace, permit to conveniently select online these decays. The fully reconstructed final state provides a strong discrimination against background. This is further enhanced by the presence of two narrow intermediate resonances (J/ψ and ϕ) whose masses can be used to impose constraints to reduce background. The high multiplicity of tracks in the final state allows a precise determination of the decay vertex position, which is crucial in the study of time evolution.

The measurement of the mixing phase $\beta_s^{J/\psi\phi}$ is conceptually similar to the measurement of the phase $\beta = \arg[(V_{td}V_{tb}^\star)/(V_{cd}V_{cb}^\star)]$ in $B^0 \to J/\psi K_S$ decays, but affected by significant additional experimental difficulties.

On average, a B_s^0 meson oscillates four times before decaying, a rate about 15 times faster than the B^0 rate. Decay-time resolution is therefore crucial to perform the analysis. The decay-time resolution depends on the relative uncertainty on the decay-length determination. The decay-length absolute uncertainty is controlled in CDF by employing silicon detectors (Sect. 3.2.2). The decay-length value depends on the lifetime of B_s^0 mesons and their boost. At CDF the average production momentum of the B_s^0 mesons is about 5 GeV/c, which yields considerably higher boost than the average boost of B_s^0 mesons produced at the B-factories that run at the $\Upsilon(5S)$ center of mass energy. This results in B_s^0 mesons to fly a significant distance before decaying, which allows a precise determination of the decay-time.

Another complication with respect to the measurement of β is the presence of decay amplitudes with different CP-eigenstates due to the spin composition of the final states. An angular analysis is required to statistically separate the various components and enhance sensitivity to the mixing phase. An accurate description of the detector effects on reconstructed particle angular distribution is needed.

As in the $B^0 \to J/\psi K_S$ case, sensitivity to the phase is obtained by separately tracking the time evolution of the initially produced B_s^0 and $\bar{B}_s^0$ mesons using flavor-tagging. Development and calibration of flavor tagging algorithms is particularly challenging in hadron collider experiments, because of large QCD backgrounds and complicated event topologies. The $\mathcal{O}(5\,\%)$ total tagging power at hadron colliders is low, compared to the $\mathcal{O}(30\,\%)$ tagging power at the B factories. On the other hand, an advantage of the $\beta_s^{J/\psi\phi}$ measurement over the β measurement results from the non-zero value of the decay-width difference $\Delta\Gamma_s$. This provides sensitivity to the mixing phase also from non flavor tagged decays, enhancing the statistical power of the event sample.

Table 2.7 Comparison of key experimental parameters

Parameter	LHCb (340 pb^{-1}) [49]	D0 (8 fb^{-1}) [48]	CDF (5.2 fb^{-1}) [30]	ATLAS (4.9 fb^{-1}) [50]
$\sigma_t(B_s^0)$ [fs]	≈50	≈100	≈90	≈100
$\sigma_m(B_s^0)$ [MeV/c^2]	≈7	≈30	≈10	≈7
Effective tagging power	≈2.1 %	≈2 %	≈4.7 %	–
Signal yield	8,300 ($t > 0.3$ ps)	5,600	6,500	22,700
S/B at peak	33/1 ($t > 0.3$ ps)	1/3	2/1	3/1

The parameters $\sigma_t(B_s^0)$ and $\sigma_m(B_s^0)$ are the mean resolutions on the measurement of the B_s^0 decay time and mass, respectively. The effective tagging power is the measurement of the capability to distinguish the production of a B_s^0 from a $\bar{B}_s^0$ meson. The symbol S/B stands for the signal to background ratio

Table 2.7 reports a comparison of key experimental parameters of the current experiments.

2.7 Analysis Strategy

The measurement of the phase $\beta_s^{J/\psi\phi}$ relies on an analysis of the time-evolution of the $B_s^0 \rightarrow J/\psi\phi$ decay in which decays from mesons produced as B_s^0 or $\bar{B}_s^0$ are studied independently, and the CP-parity of the final state is statistically determined using angular distributions. The data analysis can be dissected in four main steps: (1) selection and reconstruction of the signal event sample; (2) preparation of the analysis tools; (3) fit to the time-evolution; (4) statistical procedure to extract results and uncertainties.

The $B_s^0 \rightarrow J/\psi(\rightarrow \mu^+\mu^-)\phi(\rightarrow K^+K^-)$ event sample is collected by the CDF dimuon online-event selection system [51], which select events enriched in J/ψ decays. In the analysis, these are associated with pairs of tracks consistent with $\phi \rightarrow K^+K^-$ decays through a kinematic fit to a common space-point. A total sample corresponding to 10 fb^{-1} of data, along with a $p\bar{p} \rightarrow b\bar{b}$ production cross-section of a few tenths of microbarns, provides an event sample of few thousand B_s^0 decays. Precise momentum and vertex reconstruction, along with particle identification capabilities combined in a multivariate selection based on a machine-learning discriminator ensure isolation of an abundant and prominent signal.

Signal contributions in the $B_s^0 \rightarrow J/\psi K^+K^-$ final state other than $B_s^0 \rightarrow J/\psi\phi$ signal itself are taken into account assuming an S-wave state for the KK system.

The sensitivity to $\beta_s^{J/\psi\phi}$ is improved by fitting separately the time evolution of mesons produced as B_s^0 from those produced as $\bar{B}_s^0$ and decays occurring in different angular momenta combinations. Hence, prior to fitting the time-evolution of the decays, the algorithms that identify the b quark content of the strange-bottom meson at production are calibrated on control samples of data. Two algorithms are used for flavor tagging. One of them, the opposite side tagging algorithm has been recalibrated using data corresponding to the final data set. Similarly, the tools for the angular analysis including modeling of detector and selection sculpting are prepared.

The angular, flavor-tagging and decay-time informations are combined in an unbinned multidimensional likelihood fit, the core of the analysis, which uses information from mass, mass uncertainty, decay time, decay-time uncertainty, three-dimensional angular distributions between decay products, and outcome of the flavor tagging algorithms, to extract all the interesting physical parameters, including the phase, the width-difference, the average B_s^0 lifetime, the magnitude of the different polarization amplitudes, and a number of technical parameters of lesser importance. The complexity of the fit and some irreducible symmetries of the likelihood make the extraction of proper confidence intervals challenging from the simple fit results. Thorough simulation-based calculations are needed to construct proper confidence regions and finally extract the results.

References

1. D.H. Perkins, *Introduction to High Energy Physics*, 4th edn. (Cambridge University Press, Cambridge, 2000)
2. G. Aad et al. (ATLAS Collaboration), Observation of a new particle in the search for the Standard Model Higgs boson with the ATLAS detector at the LHC. Phys. Lett. **B716**, 1 (2012) arXiv:1207.7214 [hep-ex]
3. S. Chatrchyan et al. (CMS Collaboration), Observation of a new boson at a mass of 125 GeV with the CMS experiment at the LHC. Phys. Lett. **B716**, 30 (2012) arXiv:1207.7235 [hep-ex]
4. R. Barbieri, ICHEP2012 Physics Highlights, Contribution to the 36th International Conference on High energy Physics, Melbourn, 4–11 July 2012 arXiv:1212.3440 [hep-ph]
5. C. Quigg, Unanswered questions in the electroweak theory. Ann. Rev. Nucl. Part. Sci. **59**, 505 (2009) arXiv:0905.3187 [hep-ph]
6. M.S. Chanowitz, Electroweak symmetry breaking: unitarity, dynamics, experimental prospects. Ann. Rev. Nucl. Part. Sci. **38**, 323 (1988)
7. J. Beringer et al. (Particle Data Group), Review of particle physics. RPP. Phys. Rev. **D86**, 010001 (2012)
8. G. Isidori et al., Flavor physics constraints for physics beyond the standard model. Ann. Rev. Nucl. Part. Sci. **60**, 355 (2010) arXiv:1002.0900 [hep-ph]
9. M. Ciuchini, A. Stocchi, Physics opportunities at the next generation of precision flavor physics. Ann. Rev. Nucl. Part. Sci. **61**, 491 (2011) arXiv:1110.3920 [hep-ph]
10. J. Charles et al. (CKMfitter Group), CP violation and the CKM matrix: assessing the impact of the asymmetric B factories. Eur. Phys. J. **C41**, 1 (2005) hep-ph/0406184. Updated results and plots available at http://ckmfitter.in2p3.fr
11. S. Glashow, J. Iliopoulos, L. Maiani, Weak interactions with Lepton-Hadron symmetry. Phys. Rev. **D2**, 1285 (1970)
12. J. Augustin et al. (SLAC-SP-017 Collaboration), Discovery of a narrow resonance in e+ e− Annihilation. Phys. Rev. Lett. **33**, 1406 (1974)
13. J. Aubert et al. (E598 Collaboration), Experimental observation of a heavy particle. J. Phys. Rev. Lett. **33**, 1404 (1974)

14. F. Abe et al. (CDF Collaboration), Observation of top quark production in $\bar{p}p$ collisions. Phys. Rev. Lett. **74**, 2626 (1995) hep-ex/9503002
15. S. Abachi et al. (DØ Collaboration), Search for high mass top quark production in $p\bar{p}$ collisions at $\sqrt{s} = 1.8$ TeV. Phys. Rev. Lett. **74**, 2422 (1995) hep-ex/9411001
16. H. Albrecht et al. (ARGUS Collaboration), Observation of B0- anti-B0 mixing. Phys. Lett. **B192**, 245 (1987)
17. N. Cabibbo, Unitary symmetry and leptonic decays. Phys. Rev. Lett. **10**, 531 (1963)
18. M. Kobayashi, T. Maskawa, CP violation in the renormalizable theory of weak interaction. Prog. Theor. Phys. **49**, 652 (1973)
19. C. Jarlskog, Commutator of the quark mass matrices in the standard electroweak model and a measure of maximal CP violation. Phys. Rev. Lett. **55**, 1039 (1985)
20. L. Wolfenstein, Parametrization of the Kobayashi-Maskawa matrix. Phys. Rev. Lett. **51**, 1945 (1983)
21. A.J. Buras, M.E. Lautenbacher, G. Ostermaier, Waiting for the top quark mass, $K^+ \to \pi^+\nu\bar{\nu}$, $B^0_{(s)} - \overline{B}^0_{(s)}$ mixing and CP asymmetries in B decays. Phys. Rev. **D50**, 3433 (1994) hep-ph/9403384
22. I. Dunietz et al., In pursuit of new physics with B_s decays. Phys. Rev. D. **63**, 114015 (2001) arXiv:hep-ph/0012219 [hep-ph]
23. P. Ball, R. Fleischer, An analysis of B_s decays in the left-right-symmetric model with spontaneous CP violation. Phys. Lett. B **475**, 111 (2000)
24. I. Dunietz, $B_s - \bar{B}_s$ mixing, CP violation and extraction of CKM phases from untagged B_s data samples. Phys. Rev. **D52**, 3048 (1995) arXiv:hep-ph/9501287 [hep-ph]
25. Y. Amhis et al. (Heavy Flavor Averaging Group), Averages of b-hadron, c-hadron, and tau-lepton properties as of early 2012 (2012) arXiv:1207.1158 [hep-ex]. And updates at http://www.slac.stanford.edu/xorg/hfag
26. A. Lenz, Theoretical update of B-mixing and lifetimes (2012) arXiv:1205.1444 [hep-ph]
27. B. Bhattacharya, A. Datta, D. London, Reducing penguin pollution. Int. J. Mod. Phys. **A28**, 1350063 (2013) arXiv:1209.1413 [hep-ph]
28. X. Liu, W. Wang, Y. Xie, Penguin pollution in $B \to J/\psi V$ decays and impact on the extraction of the $B_s - \bar{B}_s$ mixing phase (2013) arXiv:1309.0313 [hep-ph]
29. S. Faller, R. Fleischer, T. Mannel, Precision physics with $B^0_s \to J/\psi\phi$ at the LHC: the quest for new physics. Phys. Rev. **D79**, 014005 (2009) arXiv:0810.4248 [hep-ph]
30. T. Aaltonen et al. (CDF Collaboration), Measurement of the CP-violating phase $\beta_s^{J/\psi\phi}$ in $B^0_s \to J/\psi\phi$ decays with the CDF II detector. Phys. Rev. **D85**, 072002 (2012) arXiv:1112.1726 [hep-ex]
31. A. Datta et al., New physics in $b \to s$ transitions and the $B^0_{d,s} \to V_1 V_2$ angular analysis. Phys. Rev. **D86**, 076011 (2012) arXiv:1207.4495 [hep-ph]
32. S. Stone, L. Zhang, S-waves and the measurement of CP violating phases in B^0_s decays. Phys. Rev. **D79**, 074024 (2009) arXiv:0812.2832 [hep-ph]
33. S. T'Jampens, Etude de la violation de la symetrie CP dans les canaux charmonium-$K^\star(892)$ par une analyse angulaire complete dependante du temps. Ph.D. thesis, University of Paris VI and VII, 2003, Available as *BABAR* thesis THESIS-03/016
34. F. Azfar et al., Formulae for the analysis of the flavor-tagged decay $B^0_s \to J/\psi\phi$. JHEP **1011**, 158 (2010) arXiv:1008.4283 [hep-ph]
35. A. Abulencia et al. (CDF Collaboration), Observation of $B^0_s - \bar{B}^0_s$ oscillations. Phys. Rev. Lett. **97**, 242003 (2006) arXiv:hep-ex/0609040 [hep-ex]
36. M. Beneke et al., Next-to-leading order QCD corrections to the lifetime difference of B(s) mesons. Phys. Lett. **B459**, 631 (1999) arXiv:hep-ph/9808385 [hep-ph]
37. M. Ciuchini et al., Lifetime differences and CP violation parameters of neutral B mesons at the next-to-leading order in QCD. JHEP **0308**, 031 (2003) arXiv:hep-ph/0308029 [hep-ph]
38. M. Beneke, G. Buchalla, I. Dunietz, Width difference in the $B_s - \bar{B}_s$ system. Phys. Rev. **D54**, 4419 (1996) arXiv:hep-ph/9605259 [hep-ph]
39. M. Beneke, G. Buchalla, A. Lenz, U. Nierste, CP asymmetry in flavor specific B decays beyond leading logarithms. Phys. Lett. **B576**, 173 (2003) arXiv:hep-ph/0307344 [hep-ph]

40. J. Charles et al., Predictions of selected flavour observables within the standard model. Phys. Rev. **D84**, 033005 (2011) arXiv:1106.4041 [hep-ph]
41. A. Lenz, U. Nierste, Theoretical update of $B_s - \bar{B}_s$ mixing. JHEP **0706**, 072 (2007) arXiv:hep-ph/0612167 [hep-ph]
42. V. Abazov et al. (D0 Collaboration), Search for CP violation in semileptonic B_s decays. Phys. Rev. **D82**, 012003 (2010) arXiv:0904.3907 [hep-ex]
43. V. M. Abazov et al. (D0 Collaboration), Measurement of the anomalous like-sign dimuon charge asymmetry with 9 fb^{-1} of $p\bar{p}$ collisions. Phys. Rev. **D84**, 052007 (2011) arXiv:1106.6308 [hep-ex]
44. R. Aaij et al. (LHCb Collaboration), Measurement of the $B_s^0 - \bar{B}_s^0$ oscillation frequency Δm_s in $B_s^0 \to D_s^-(3)\pi$ decays. Phys. Lett. **B709**, 177 (2012) arXiv:1112.4311 [hep-ex]
45. T. Aaltonen et al. (CDF Collaboration), First flavor-tagged determination of bounds on mixing-induced CP violation in $B_s^0 \to J/\psi\phi$ decays. Phys. Rev. Lett. **100**, 161802 (2008) arXiv:0712.2397 [hep-ex]
46. V. Abazov et al. (D0 Collaboration), Measurement of B_s^0 mixing parameters from the flavor-tagged decay $B_s^0 \to J/\psi\phi$. Phys. Rev. Lett. **101**, 241801 (2008) arXiv:0802.2255 [hep-ex]
47. The CDF and D0 Collaborations, Combination of D0 and CDF results on $\Delta\Gamma_s$ and the CP-violating phase $\beta_s^{J/\psi\phi}$, CDF Note 9787, D0 Note 5928-CONF (2009)
48. V. Abazov et al. (D0 Collaboration), Measurement of the CP-violating phase $\phi_s^{J\psi\phi}$ using the flavor-tagged decay $B_s^0 \to J\psi\phi$ in 8 fb^{-1} of $p\bar{p}$ collisions. Phys. Rev. **D85**, 032006 (2012) arXiv:1109.3166 [hep-ex]
49. R. Aaij et al. (LHCb Collaboration), Measurement of the CP-violating phase ϕ_s in the decay $B_s^0 \to J/\psi\phi$. Phys. Rev. Lett. **108**, 101803 (2012) arXiv:1112.3183 [hep-ex]
50. G. Aad et al. (ATLAS Collaboration), Time-dependent angular analysis of the decay $B_s^0 \to J/\psi\phi$ and extraction of $\Delta\Gamma_s$ and the CP-violating weak phase ϕ_s by ATLAS (2012) arXiv:1208.0572 [hep-ex]
51. D. Acosta et al. (CDF Collaboration), Measurement of b hadron masses in exclusive J/ψ decays with the CDF detector. Phys. Rev. Lett. **96**, 202001 (2006) hep-ex/0508022

Chapter 3
The Collider Detector at the Fermilab Tevatron

This chapter provides a concise description of the Tevatron collider and of the CDF II detector. More details are provided on the tracking and muon detector systems, on account of their importance in the present analysis.

3.1 The Fermilab Tevatron Collider

The Tevatron collider, an underground circular proton synchrotron, was the last stage of a system of accelerators, storage rings, and transfer lines, located at the Fermi National Accelerator Laboratory (FNAL or Fermilab), about 50 km west from Chicago, in the United States (Fig. 3.1).

The Tevatron has been the world highest-energy accelerator since 1985, when it started operation as the first superconducting proton-synchrotron, until 2007, when the CERN Large Hadron Collider started operating. In its last period of collider operations, called *Run II*, it provided collisions of antiprotons with protons at a center-of-mass energy of 1.96 TeV. It employed about 1,000 dipole bending magnets with niobium-titanium superconducting coils in a 1 km radius ring. Each dipole magnet was 6.4 m long and was cooled with liquid helium down to 4.3 K. The dipole field reached 4.2 T. When in collider mode, "bunches" of protons spaced by 396 ns collided against a similar beam of antiprotons.

In the two interaction points, conventionally named B0 and D0, the colliding beams were shrunk in the plane transverse to the beam to a diameter of approximate Gaussian shape with about 32 μm width.

The performance of the Tevatron collider is defined in terms of two key parameters: the center-of-mass energy, $\sqrt{s}$, and the instantaneous luminosity, $\mathscr{L}$, that is the coefficient of proportionality between the rate of a given process and its cross-section σ,

$$\frac{dN}{dt} \ [\text{events/s}] = \mathscr{L} \left[\text{cm}^{-2}\text{s}^{-1} \right] \sigma \left[\text{cm}^2 \right].$$

© Springer International Publishing Switzerland 2015

S. Leo, *CP Violation in $B_s^0 \to J/\psi\phi$ Decays*, Springer Theses,

DOI 10.1007/978-3-319-07929-5_3

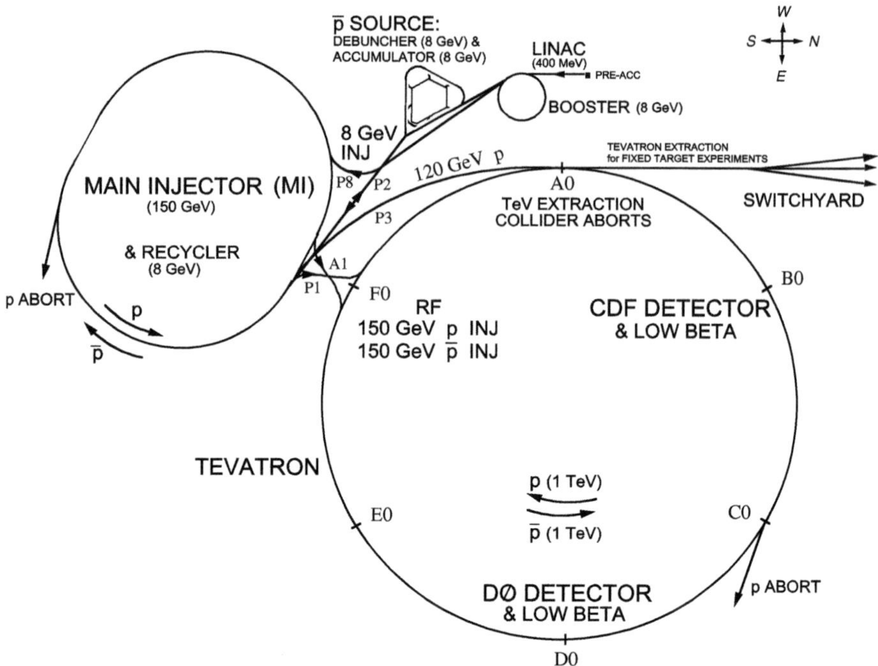

Fig. 3.1 Sketch of the Fermilab's accelerators system

The time-integral of the luminosity (integrated luminosity) is therefore a measure of the average expected number of events, N, produced in a finite time T,

$$N(T) = \int_0^T \mathscr{L}\, \sigma \, dt.$$

Assuming an ideal head-on $p\bar{p}$ collision with no crossing angle between the beams, the instantaneous luminosity at the Tevatron was defined as

$$\mathscr{L} = 10^{-5} \frac{N_p N_{\bar{p}} B f \beta\gamma}{2\pi\beta^{\star}\sqrt{(\epsilon_p + \epsilon_{\bar{p}})_x (\epsilon_p + \epsilon_{\bar{p}})_y}}\, F\left(\sigma_z/\beta^{\star}\right) \quad \left[10^{30}\mathrm{cm}^{-2}\mathrm{s}^{-1}\right],$$

where N_p ($N_{\bar{p}}$) is the average number of protons (antiprotons) in each bunch ($N_p \approx 8.1 \times 10^{11}$ and $N_{\bar{p}} \approx 2.9 \times 10^{11}$), B (36) is the number of bunches per beam circulating into the ring, f (47.713 kHz) is the revolution frequency, $\beta\gamma$ is the relativistic factor of the Lorentz boost (1,045.8 at 980 GeV), F is an empiric form factor which depends on the ratio between the longitudinal width of the bunch ($\sigma_z \approx 60\,\mathrm{cm}$) and the "betatron function" calculated at the interaction point ($\beta^{\star} \approx 31\,\mathrm{cm}$), and finally ϵ_p

($\epsilon_{\bar{p}}$) is the 95 % normalized emittance of the proton (antiproton) beam ($\epsilon_p \approx 18\pi$ mm $\times$ mrad and $\epsilon_{\bar{p}} \approx 13\pi$ mm $\times$ mrad after injection).[1]

The most important factor determining the luminosity was the antiproton current that could be efficiently transferred through the accelerator chain for final collisions. The particles were accelerated in bunches enclosed in radiofrequency (RF) buckets. A bucket is one interval of the longitudinal restoring force provided by the RF cavities that resulted in a stable phase-space where a bunch may be captured and accelerated. During the acceleration process, bunch emittance was reduced (cooling) and groups of adjacent bunches ("trains") were eventually stored and accelerated to maximum energy in the Tevatron. During a continuous period of collisions (*"store"*), which could last up to 24 h, the Tevatron injector chain provided beams for a number of fixed-target experiments (primarily on neutrino beams). The procedure for obtaining a *store* is described in the following subsections. Further details can be found in Refs. [1, 2].

3.1.1 Proton Production

Protons were produced from gaseous hydrogen H_2, which was negatively ionized to allow an essentially loss-free acceleration to a 750 keV Cockroft-Walton accelerator.

The 750 keV H^- ions were then accelerated up to 400 MeV by a 130 m long Alvarez-type linear accelerator (*Linac*, Fig. 3.1). The H^- beam pulse lasted typically 20 ms and was injected into a synchrotron called *Booster*. When entering the Booster, H^- ions passed through a carbon foil where the two electrons were removed. The Booster had a circumference of 475 m and accelerated protons from 400 MeV to 8 GeV compacting them into bunches of about 5×10^{12} particles each. The bunches were then transfered into the *Main Injector*, a synchrotron which increased their energy up to 150 GeV, and finally into the Tevatron where superconducting magnets kept them on an approximately circular orbit while the antiproton beam was injected.

Injecting H^- ions rather than protons into the Booster allowed the injection to proceed over multiple revolutions of the beam around the Booster ring (usually 10–12). If protons were to be injected, the magnetic field used to inject new protons onto orbit in the Booster would have deflected the already revolving protons out of orbit.

---

[1] The form factor F is a parameterization of the longitudinal profile of the beams in the collision region, which assumes the characteristic shape of an horizontal "hourglass" centered at the interaction point. The betatron function is a parameter convenient for solving the equation of motion of a particle through an arbitrary beam transport system; $\beta^\star$ is a local function of the magnetic properties of the ring and it is independent of the accelerating particle. The emittance ϵ measures the phase space occupied by the particles of the beam; three independent two-dimensional emittances are defined, for each of them $\sqrt{\beta^\star \epsilon}$ is proportional to the statistical width of the beam in the corresponding phase plane.

3.1.2 Antiproton Production and Accumulation

The radiofrequency-bunched proton beam was extracted from the Main Injector at 120 GeV and brought to collide against a 7 cm-thick rotating nickel target, where many secondary particles, including antiprotons, were produced. These were focused by a lithium lens and analyzed in a magnetic spectrometer that selected negatively-charged particles. Antiprotons were produced over a wide momentum range, with a broad maximum around 8 GeV and an efficiency of about 2×10^{-5} per interacting proton.[2] The bunched antiproton beam was accepted with a momentum spread of about 2.5 % by a *Debuncher* synchrotron (Fig. 3.1) where, by radiofrequency manipulation, it was turned into a continuous, nearly monochromatic 8 GeV beam. The antiproton debunched-beam was transferred to the *Accumulator* Ring, housed in the same tunnel of the Debuncher, which was a triangle-shaped storage ring and collected pulses from the Debuncher aver many hours. In the Accumulator, a higher-intensity antiproton beam was stored, owing to its larger acceptance. In both the Debuncher and the Accumulator, the longitudinal and transverse momentum spread of the beam was reduced ("*cooled*") by stochastic cooling.[3] Since 2004, optimized antiproton accumulation was achieved using the *Recycler Ring* (Fig. 3.1). This was a constant 8 GeV-energy storage-ring placed in the Main Injector enclosure, that used permanently magnetized strontium ferrite. It was used to gather antiprotons that were periodically transferred from the Accumulator (with $\approx$95 % transfer efficiency) thus maintaining it at its optimum intensity regime. Relativistic electron cooling was also successfully implemented in the Recycler, further enhancing the Tevatron performance [4].[4]

3.1.3 Injection and Collisions

Every approximately 10–20 h, enough antiprotons had accumulated for a high luminosity store. Accumulation was thus stopped in preparation for injection. A set of seven proton bunches was extracted from the Booster, injected into the Main Injector, accelerated to 150 GeV, coalesced with about 90 % efficiency into a single bunch of

---

[2] Typically, 21 antiprotons were collected for each 106 protons on target, resulting in a stacking rate of approximately 10–20 mA/h.

[3] Stochastic cooling is a technique used to reduce the transverse momentum and energy spread of a beam without beam loss. This is achieved by applying iteratively a feedback mechanism that senses the beam deviation from the ideal orbit with electrostatic plates, processes and amplifies it, and transmits an adequately-sized synchronized correction pulse to another set of plates downstream [3]. Bunch rotation is an RF manipulation technique that, using adequate phasing, transforms a beam with a large time spread and a small energy spread in a beam with a large energy spread and a small time spread, or viceversa.

[4] Electron cooling is a method of damping the transverse motion of the antiproton beam through the interaction with an electron beam propagating together at the same average velocity.

approximately 3×10^{12} protons, and then injected into the Tevatron.[5] This process was repeated every 12.5 s, until 36 proton bunches, separated by 396 ns, were loaded into the Tevatron central orbit. Typically, 65 % of the protons in the Main Injector were successfully transferred to the Tevatron. Since protons and antiprotons circulated in the same enclosure, sharing magnet and vacuum systems, electrostatic separators (about 30 pairs of metal plates) were activated to keep protons and antiprotons into two non-intersecting closed helical orbits.

Four sets of 7–11 antiproton bunches were extracted from the Accumulator (or from the Recycler) to the Main Injector, accelerated to 150 GeV, coalesced with about 80 % efficiency into four 8×10^{11} antiproton bunches separated by 396 ns, and then injected into the Tevatron, where protons were counter-rotating. The injection process was repeated nine times until 36 antiproton bunches circulated in the Tevatron.

The energy of the machine was then increased in about 10 s from 150 to 980 GeV at which energy one particle completes the full revolution of the Tevatron circumference in 21 μs at $0.9999996c$, and collisions began at the two interaction points, D0 and B0 (where the D0 and the CDF II detectors were respectively located). Special quadrupole magnets (*low-β squeezers*) located at both extremities of the detectors along the beam pipe "squeezed" the beam in the transverse direction to maximize the luminosity inside the detectors. Then the beam transverse-profile was shaped to its optimized configuration by mean of iron plates which acted as collimators and trimmed the transverse beam-halo. The interaction region had a roughly Gaussian distribution in both transverse ($\sigma_T \approx 30\,\mu\mathrm{m}$) and longitudinal ($\sigma_z \approx 28\,\mathrm{cm}$) planes approximately centered in the nominal interaction point. When the beam profile was narrow enough and the conditions were safely stable, the detectors were powered and data taking started.

During collisions, the luminosity decreased exponentially from its peak initial value of typically $3.2 \times 10^{32}\,\mathrm{cm}^{-2}\mathrm{s}^{-1}$, because of the beam-gas and beam-halo interactions. The decrease is about a factor of 3 (5) for a store of ≈ 10 (20) h. In the meantime, antiproton production and storage continued. When the antiproton stack was sufficiently large (about 4×10^{12} antiprotons) and the circulating beams were degraded, the detector high-voltages were switched off and the store was dumped. The beam was extracted via a switch-yard and sent to an absorption zone. Beam abortion could occur also accidentally when the temperature of a superconducting magnet fluctuated above the critical value and a magnet *quenched*, destroying the orbit of the beams. The time between the end of a store and the beginning of collisions of the next one was typically 1 h, during which calibrations of the sub-detectors and test runs with cosmics were performed.

---

[5] Coalescing is the process of compacting into one dense bunch many smaller bunches.

3.1.4 Run II Performances and Achievements

The last period of Tevatron operations, started in March 2001 and continued through September 2011, is commonly referred to as Run II.

At the end of the Run II, typical Tevatron luminosities were constantly well above $3.2 \times 10^{32}\,\mathrm{cm}^{-2}\mathrm{s}^{-1}$, with a record peak of $4.4 \times 10^{32}\,\mathrm{cm}^{-2}\mathrm{s}^{-1}$.

The integrated luminosity delivered by the Tevatron has greatly progressed over the years, reaching a total of 11 fb^{-1} from $p\bar{p}$ collisions delivered to either experiment, thus enabling CDF and D0 to carefully study the standard model and characterize many of its important features for the first time.

3.2 The CDF II Detector

The CDF II detector was a large multi-purpose solenoidal magnetic spectrometer surrounded by full coverage, projective-geometry calorimeters and fine-grained muon detectors installed at the B0 interaction point of the Tevatron (Fig. 3.2). The CDF II

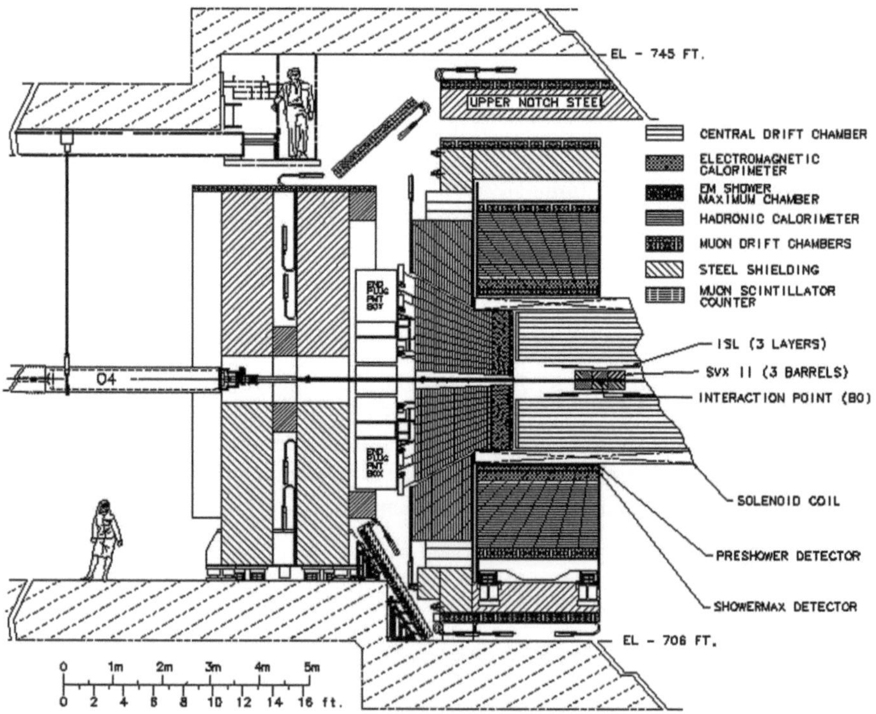

Fig. 3.2 Elevation view of one half of the CDF Run II detector. The TOF and the small-angle detectors are not depicted

detector was designed and constructed with an approximately cylindrically symmetric layout both in the azimuthal plane and in the "forward" ($z > 0$, east) "backward" ($z < 0$, west) directions. It comprised a number of coaxial sub-detectors that provided information to determine energy, momentum, and nature of a broad range of particles produced in 1.96 TeV $p\bar{p}$ collisions:

- a charged-particle tracking system composed by three silicon microstrip detectors (L00, SVX II and ISL, from inner to outer radii) and an open-cell drift chamber (COT) housed inside a superconducting solenoid providing a 1.4 T axial magnetic field;
- a time of flight detector, radially outside the COT for identification of charged particles with transverse momenta of less than 1.5 GeV;
- calorimeters, located outside the magnet and used to measure the energy of electrons, photons, and hadron jets and to provide a muon filter;
- dedicated outermost detectors to identify penetrating muons;
- two small-angle spectrometers in the very forward and backward regions with respect to the main detector for specialized studies of diffraction processes;
- luminosity monitors.

A detailed description of the CDF II detector can be found in Ref. [5]. In the following we outline the general features of the subsystems most relevant for this work.

3.2.1 Coordinates and Notation

CDF II employed a right-handed Cartesian coordinates system with the origin in the B0 interaction point, assumed coincident with the center of the drift chamber. The positive z axis lay along the nominal beam-line pointing toward the proton direction. The (x, y) plane was therefore perpendicular to either beams, with positive y axis pointing vertically upward and positive x axis in the horizontal plane of the Tevatron, pointing radially outward with respect to the center of the ring.

Since the colliding beams of the Tevatron were unpolarized, the resulting physical observations are invariant under rotations around the beam line axis. A cylindrical (r, ϕ, z) coordinates system is particularly convenient to describe the detector geometry. Throughout this thesis, *longitudinal* (or *axial*) means parallel to the proton beam direction (i.e., to the z axis) and *transverse* means perpendicular to the proton direction (i.e., in the (x, y) or (r, ϕ) plane).

Since the protons and antiprotons are composite particles, the actual interaction occurred between their component partons (valence or sea quarks and gluons). Each parton carried a varying fraction of the (anti)proton momentum, not known on a event-by-event basis. As a consequence of the possible imbalance in the longitudinal components of the momenta of interacting partons, possible large velocities along z for the center-of-mass of the parton-level interaction could occur. In hadron collisions, it is customary to use *rapidity* as a variable invariant under z boosts and unit of relativistic phase-space, instead of the polar angle θ:

$$Y = \frac{1}{2} \ln \left[\frac{E + p \cos \theta}{E - p \cos \theta} \right], \tag{3.1}$$

where (E, p) is the energy-momentum four-vector of the particle.[6] However, a measurement of rapidity requires a detector with full particle identification capability because of the mass term entering E. Thus Y is replaced with its ultrarelativistic approximation η, usually valid for products of high-energy collisions except at the most forward angles,

$$Y \xrightarrow{p \gg m} \eta + \mathcal{O}(m^2/p^2), \tag{3.2}$$

where the *pseudorapidity* $\eta \equiv -\ln\left[\tan(\theta/2)\right]$ is only a function of the polar angle. As the event-by-event longitudinal position of the interaction was distributed around the nominal interaction point with 30 cm rms width, it is useful to distinguish *detector pseudorapidity*, η_{det}, measured with respect to the $(0, 0, 0)$ nominal interaction point, from *event pseudorapidity*, η, which is measured with respect to the z_0 position of the event vertex where the particle originated.[7]

Mapping the solid angle in terms of (pseudo)rapidity and azimuthal angle is also convenient because the density of final-state particles in energetic hadronic collisions is approximately flat in the (Y, ϕ) space. Other convenient variables used are the transverse component of the momentum with respect to the beam axis (p_T), the "transverse energy" (E_T), and the approximately Lorentz-invariant distance in the $\eta - \phi$ space ΔR, respectively defined as

$$p_T \equiv (p_x, p_y) \rightarrow p_T \equiv p \sin(\theta), \quad E_T \equiv E \sin(\theta), \quad \text{and} \quad \Delta R \equiv \sqrt{\Delta\eta^2 + \Delta\phi^2}. \tag{3.3}$$

3.2.2 Tracking System

Three-dimensional charged particle tracking was achieved through an integrated system consisting of three silicon inner subdetectors and a large outer drift-chamber, all contained in a superconducting solenoid. The 1.4 T magnetic field and the 136 cm total lever arm provided excellent tracking performances.

Charged particles left small ionization energy depositions as they passed through the layers of the tracking system. Using a set of spatial measurements of these depositions ("hits"), pattern recognition and kinematic fitting algorithms reconstructed the particle's original trajectory measuring the parameters of the helicoidal trajectory

---

[6] The rapidity can be derived from the Lorentz-invariant cross-section: $E\frac{d^3\sigma}{(dp)^3} = E\frac{d^2\sigma}{\pi p_T dp_T dp_z}$. Observing that only E and p_z change under z boosts, we can replace them by a variable Y such as $E\frac{dY}{dp_z} = 1$. Solving for Y we get Eq. (3.1).

[7] An idea of the difference is given by considering that $|\eta_{\text{det}} - \eta_{part}| \approx 0.2$ if the particle is produced at $z = 60$ cm from the nominal interaction point.

that best matched the observed path in the tracking detector. The silicon detectors provided excellent impact parameter, azimuthal angle, and z_0 resolution, while the drift chamber provided excellent resolution of the curvature and ϕ_0. Together they provided a very accurate measurements of the charged particles' trajectories, which is a key element of the present analysis, allowing a resolution on the $B_s^0 \to J/\psi\phi$ reconstructed mass of 9 MeV/c^2.

3.2.2.1 The Inner Silicon Tracker

The silicon tracking system, as shown in Fig. 3.3 in both (r, ϕ) and (r, z) projections, was composed of three approximately cylindrical coaxial subsystems: the *Layer00* (L00), the *Silicon Vertex detector* (SVX II) and the *Intermediate Silicon Layer* (ISL).

L00 [6] was the innermost subsystem and consisted of one layer of single-sided, AC-coupled, microstrip silicon sensors placed on the beam pipe at radii of 1.35 and 1.62 cm. It provided full azimuthal and $|z| \lesssim 47$ cm longitudinal coverage. Longitudinally adjacent sensors ($0.84 - 1.46$ cm $\times$ 7.84 cm) were ganged in modules of 15.7 cm active-length arranged into twelve partially-overlapping ϕ sectors, and six longitudinal barrels. The strips were parallel to the beam axis allowing sampling of tracks in the (r, ϕ) plane. The inter-strip implant pitch of 25 μm with floating alternate strips resulted in 50 μm readout pitch. The analog signals of the 13,824 channels were fed via fine-pitch cables, about 50 cm long, to the front-end electronics outside the tracking volume.

The core of the silicon tracker was SVX II [7]. The SVX II was a fine-resolution silicon micro-strip vertex detector that provided five three-dimensional samplings of tracks at 2.45, 4.1, 6.5, 8.2 and 10.1 cm (or, depending on the ϕ sector, at 2.5,

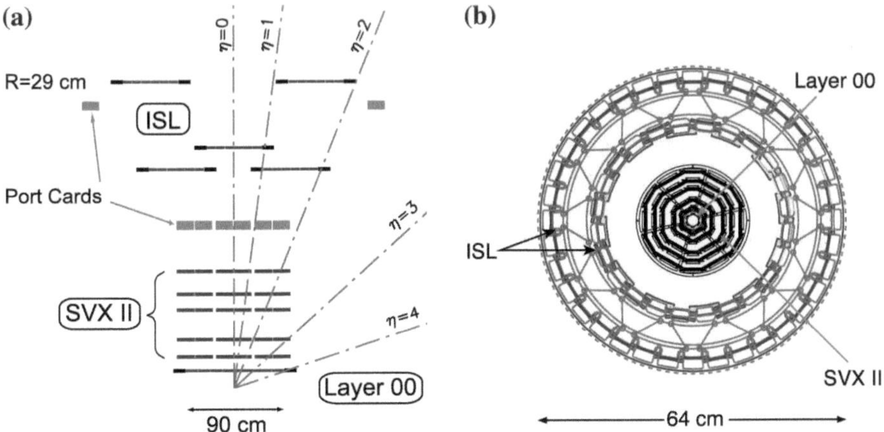

Fig. 3.3 View of the CDF II silicon system, including the SVX II cooling bulkheads and ISL support structure, in the (r, z) (**a**) and (r, ϕ) (**b**) planes. The z scale is highly compressed

4.6, 7.1, 8.7 and 10.6 cm) of radial distance from the beam with full pseudorapidity coverage in the $|\eta_{\text{det}}| \lesssim 2$ region. This corresponded to a length of $|z| \lesssim 96$ cm along the beam-line, sufficient to cover the longitudinal spread of the luminous region. The SVX II had a cylindrical geometry coaxial with the beam, and its mechanical layout was segmented in three 32 cm axial sections ("barrels") times twelve 30° azimuthal sectors ("wedges") times five equally-spaced radial layers. A small overlap between the edges of adjacent azimuthal sectors helped wedge-to-wedge alignment (Fig. 3.3b). Sensors in a single layer were arranged into independent longitudinal readout units.

The ISL [8] detector was placed at intermediate radial distance between the SVX II and the drift chamber and had polar coverage up to $|\eta_{\text{det}}| < 2$ and a total length of 174 cm along z. At $|\eta_{\text{det}}| \lesssim 1$, a single layer of double-readout silicon microstrip sensors was mounted on a cylindrical barrel at radius of 22.6 cm (or 23.1 cm). At $1 \lesssim |\eta_{\text{det}}| \lesssim 2$ two layers of silicon sensors were arranged into two pairs of concentric barrels (inner and outer). In the inner (outer) barrel, staggered ladders alternated at radii of 19.7 and 20.2 cm (28.6 and 29 cm). One pair of barrels was installed in the forward region, the other pair in the backward region. Each barrel was azimuthally divided into a 30° structure matching the SVX II segmentation. The basic readout unit consisted of an electronic board and three sensors ganged together resulting in a total active length of 25 cm.

The total amount of material in the silicon system, averaged over ϕ and z, varied approximately as $0.1 X_0/\sin\theta$ in the $|\eta_{\text{det}}| \lesssim 1$ region, and typically doubled in $1 \lesssim |\eta_{\text{det}}| \lesssim 2$ because of the presence of cables, cooling bulk-heads, and portions of the support frame.[8] The average amount of energy loss for a charged particle was roughly 9 MeV. The total heat load of the silicon system was approximately 4 kW. To prevent thermal expansion, relative detector motion, increased leakage-current, and chip failure due to thermal heating, the silicon detectors and the associated front-end electronic boards were held at roughly constant temperature ranging from -6 to $-10\,°C$ for L00 and SVX II, and around $10\,°C$ for ISL, by an under-pressurized water and ethylene-glycol coolant flowing in aluminum pipes integrated in the supporting structures.[9]

The resolution on the hit position for all silicon sensors was about $11\,\mu m$ in the (r, ϕ) plane, thus allowing to reach about $20\,\mu m$ resolution on the impact parameter of high-p_T tracks which degraded to about $35\,\mu m$ at 2 GeV/c. This precision provided a powerful discriminator to identify long-lived hadrons containing heavy-flavored quarks already at trigger level. The SVX II resolution in the z direction was approximately 70–$100\,\mu m$.

---

[8] The symbol X_0 indicates the radiation length.

[9] The cooling fluid is maintained under the atmospheric pressure to prevent leaks in case of damaged cooling pipes.

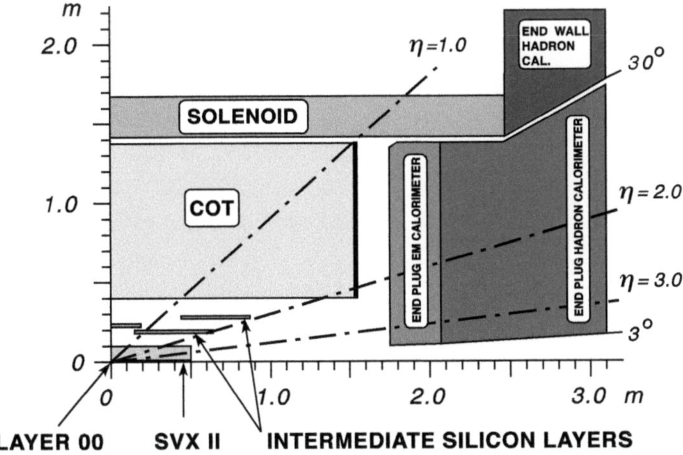

Fig. 3.4 Elevation view of one quadrant of the inner portion of the CDF II detector showing the tracking volume surrounded by the solenoid and the forward calorimeters

3.2.2.2 The Central Outer Tracker

The main tracking detector at CDF was the *central outer tracker* (COT) [9], a large multi-wire, open-cell drift chamber that provided charged-particle tracking in the central pseudorapidity region $|\eta_{det}| \lesssim 1$ (Fig. 3.4).

The COT had an hollow-cylindrical geometry, its active volume spanned from 43.4 to 132.3 cm in radius and $|z| \lesssim 155$ cm in the axial direction, and was filled with a 50:50 gas admixture of argon and ethane bubbled through isopropyl alcohol (1.7 %) that provided fast drift of ionization electrons ($\approx$100 µm/ns). Arranged radially into eight "super-layers", it contained 96 planes of wires that ran the length of the chamber between two end-plates (Fig. 3.5a). Each super-layer was divided into ϕ cells; within a cell, the trajectory of a charged particle was sampled at 12 radii (spaced 0.583 cm apart) where sense wires (anodes) were strung. Four super-layers employed sense-wires parallel to the beam axis, for the measurement of the hit coordinates in the (r, ϕ) plane. These were radially interleaved with four stereo super-layers whose wires were alternately canted at angles of $2°$ and $-2°$ with respect to the beamline. Combined readout of stereo and axial super-layers allowed the measurement of the (r, z) hit coordinates. Each super-layer was azimuthally segmented into open drift cells. Figure 3.5b shows the drift cell layout, which consisted of a wire plane closed azimuthally by cathode sheets spaced approximately 2 cm apart. The wire plane contained sense wires alternating with field-shaping wires, which controlled the gain on the sense wires thus optimizing the electric field intensity. The cathode was a 6.35 µm thick Mylar sheet with vapor-deposited gold, shared with the neighboring cell.[10] Innermost and outermost radial extremities of a cell (i.e., the boundaries

---

[10] Gold, used also for the wires, was chosen because of its good conductivity, high work function, resistance to etching by positive ions, and low chemical reactivity.

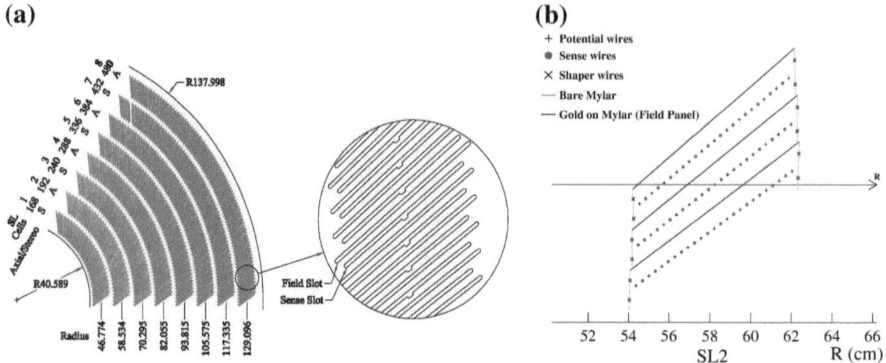

Fig. 3.5 A 1/6 section of the COT end-plate (**a**); for each super-layer the total number of cells, the wire orientation (axial or stereo), and the average radius in cm are given; the enlargement shows in details the slot where the wire planes (sense and field) are installed. Sketch of an axial cross-section of three cells in the super-layer 2 (**b**); the *arrow points* into the radial direction

between super-layers) were closed both mechanically and electrostatically by Mylar strips with an additional field-shaping wire attached, the shaper wire.

Both the field sheet and the wire plane had a center ($z \approx 0$) support rod that limited motion due to electrostatic forces. Each wire plane contained 12 sense, 13 field-shaping and 4 shaper wires, all made of $40\,\mu\mathrm{m}$ diameter gold-plated tungsten. Wire planes were not aligned with the chamber radius. A $\zeta = 35°$ azimuthal tilt partially compensated for the Lorentz angle of the drifting electrons in the magnetic field.[11] The tilted-cell geometry helped in the drift velocity calibration, since high-p_T (radial) tracks sampled the full range of drift distances within each super-layer. Further benefit of the tilt was that the left-right ambiguity was resolved for particles coming from the z axis, since the ghost track in each super-layer appeared azimuthally rotated by $\arctan[2\tan(\zeta)] \approx 54°$, simplifying pattern recognition.

The COT single-hit resolution was $140\,\mu\mathrm{m}$, including a $75\,\mu\mathrm{m}$ contribution from the ≈ 0.5 ns uncertainty on the measurement of the $p\bar{p}$ interaction time. Internal alignment of the COT cells was maintained within $10\,\mu\mathrm{m}$ using cosmic rays. Curvature effects from gravitational and electrostatic sagging were under control within $0.5\,\%$ by equalizing the difference of E/p between electrons and positrons as a function of $\cot\theta$. The total amount of material in the COT, including the gas mixture, corresponded to 0.017 radiation lengths for electrons. The COT was read by 30,240 linear electronic channels and was capable of measuring also the specific ionization energy, dE/dx providing 1.5σ separation between charged kaons and pions with

---

[11] In the presence of crossed electric ($\boldsymbol{E}$) and magnetic ($\boldsymbol{B}$) fields, electrons drifting in a gas move at an angle ζ with respect to the electric field direction, given by $\zeta \approx \arctan\left(\frac{v(E, B=0)B}{kE}\right)$, where $v(E, B = 0)$ is the drift velocity without a magnetic field, and k is a $\mathcal{O}(1)$ empirical parameter that depends on the gas and on the electric field. A common solution for this problem consists in using tilted cells (i.e. tilted drift electric field) that compensate the Lorentz angle linearizing the time-to-distance relation.

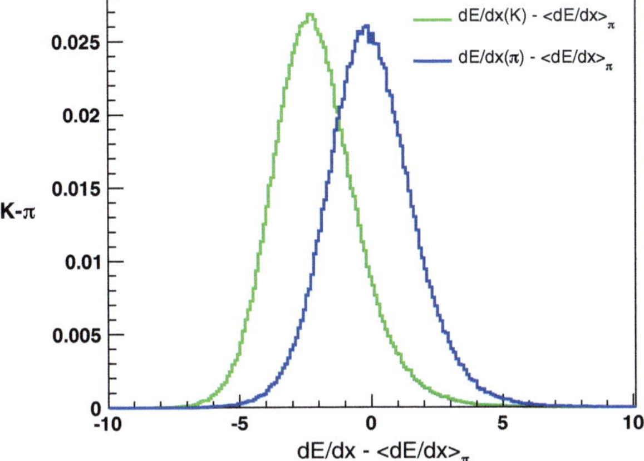

Fig. 3.6 Data distributions for dE/dx residuals

$p_T > 2$ GeV/c (Fig. 3.6). The excellent CDF II tracking performance is key element of the analysis. For our signal, $B_s^0 \rightarrow J/\psi\phi$, it results indeed in a reconstructed mass resolution of ≈ 9 GeV$/c^2$ aiming a very clean sample selection.

3.2.3 Time of Flight Detector

Between the COT and the magnet, a layer of scintillator bars measured the charged particle time of flight (TOF) from the collision point [10]. *Particle identification* (PID) was achieved combining TOF information with dE/dx resulting in a 2σ separation, assuming Gaussian distributions, of pions from kaons with momenta not exceeding 1.5 GeV Fig. 3.7.

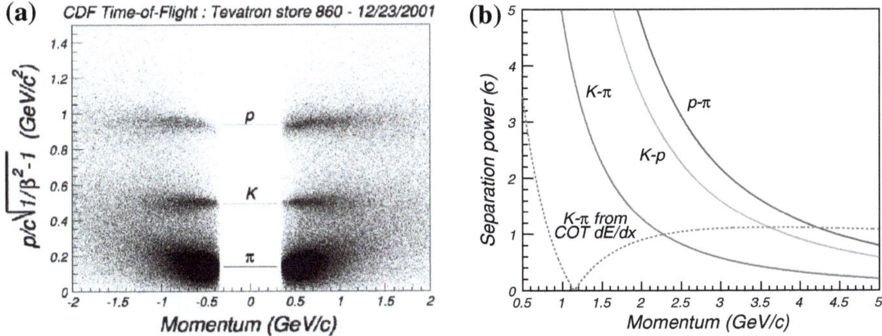

Fig. 3.7 TOF distribution of different particles (**a**) and separation power for TOF compared to that of dE/dx (**b**)

The TOF detector was composed by 216 scintillator bars, with a slightly trapezoidal cross-section of 4 cm maximum basis, 4 cm thickness and 2.79 m length. Light was collected by photomultipliers at the ends of the bars. Single hit position along the scintillator bars was determined by comparing the timing of the photomultiplier signals. The TOF time resolution was approximately 120 ps. For the TOF measurement the collision time t_0 had to be known. This was found with about 50 ps uncertainty by a fit to all tracks in the event. Unfortunately, in high luminosity conditions, the occupancy of the single bars determined a degradation in efficiency of about 50 % per track.

3.2.4 Calorimeters

Located immediately outside the solenoid, the CDF calorimeter covered a solid angle of nearly 4π around the p$\bar{\text{p}}$ interaction point up to $|\eta_{\text{det}}| \lesssim 3.6$. It measured the energy deposited by hadrons, electrons, and photons, using "shower" sampling based on layers of high-Z passive absorber interspersed with layers of plastic scintillator. The calorimeters were segmented in solid angle around the nominal collision point and segmented into two compartments radially outward from the collision point (in-depth segmentation). Angular segmentation was organized in projective *towers*. Each tower was an independent readout unit, which subtended a portion of the solid angle, namely a rectangular cell in the $(\eta_{\text{det}} - \phi)$ space, with respect to the nominal interaction point. In-depth segmentation of each tower consisted of two independent compartments: the inner one sampled the electromagnetic component of the shower, while the outer one sampled the hadronic fraction of the deposited energy. Different fractions of energy release in the two compartments were used to distinguish photons and electrons from hadrons.

Electrons and photons were identified and had their energy sampled in the EM calorimeter by a set of thin scintillator layers interspersed with lead absorbers. The EM calorimeter was subdivided into three portions: the central EM calorimeter (CEM), covering the region $|\eta| \lesssim 1.1$, and the two plug EM calorimeters (PEM), covering the forward regions $1.1 < |\eta| < 3.6$. The CEM and PEM energy resolutions were measured to be:

$$\frac{\sigma_E}{E} = \frac{13.5\,\%}{\sqrt{E_T}} \oplus 2\,\% \quad \text{and} \quad \frac{\sigma_E}{E} = \frac{16\,\%}{\sqrt{E_T}} \oplus 1\,\%. \tag{3.4}$$

The identification of hadrons and the measurement of their energy are performed by calorimeter towers located behind the EM ones: the central hadronic calorimeter (CHA), covering the region $|\eta_{\text{det}}| < 0.9$, two calorimeter rings that cover the gap between CHA and PHA in the region $0.7 < |\eta_{\text{det}}| < 1.3$, called the wall hadron calorimeters (WHA) and the two plug hadron calorimeters (PHA) covering the forward regions $1.3 < |\eta_{\text{det}}| < 3.6$. The resolutions of CHA, WHA, and PHA, from test beam measurements (response to single pions) are

$$\frac{\sigma_E}{E} \approx \frac{50\,\%}{\sqrt{E}} \oplus 3\,\%, \quad \frac{\sigma_E}{E} \approx \frac{75\,\%}{\sqrt{E}} \oplus 4\,\%, \quad \text{and} \quad \frac{\sigma_E}{E} \approx \frac{80\,\%}{\sqrt{E}} \oplus 5\,\%, \quad (3.5)$$

respectively. In this measurement, calorimeters are only used in the reconstruction of jets for identification of bottom-strange flavor at production. More details on their technology is in Ref. [5].

3.2.5 Muons Detectors

The furthest detector component from the beampipe are the muon chambers, consisting in scintillating counters and drift tubes [11, 12] installed at various radial distances from the beam. The muon detector was divided in subsystems: the central muon detector (CMU), the central muon upgrade detector (CMP), the central muon extension detector (CMX), and the intermediate muon upgrade (IMU).

3.2.5.1 CMU

The CMU detector was located outside of the central hadronic calorimeter, with a layer of steel shielding between the two to help in suppressing the fraction of hadrons decaying close to calorimeters end and producing secondary particles punching trough the muon chamber. The CMU provided coverage for $|\eta| < 0.6$. It was made up of 2,880 axially positioned single-wire proportional-chambers. It was segmented into $15°$ wedges, but gaps between the wedges decreased the total ϕ acceptance to $84\,\%$. The wedges were segmented into three smaller wedges containing an array of drift cells, 4 cells wide and 4 cells deep. A muon had to have $p_T \gtrsim 1.4$ GeV/c to reach the CMU.

3.2.5.2 CMP

The CMP was located beyond the CMU, and was separated from it by additional 60 cm of steel. It had the same $|\eta|$ coverage as the CMU, but it was built around the return yoke of the solenoid, which intruded on the ϕ coverage in places. It was composed of 1,076 axially-positioned single wire proportional chambers. A muon had to have $p_T \gtrsim 2.2$ GeV/c to reach the CMP. Muons that leave a signal in both the CMU and CMP were called CMUP muons. Such a sample had a subpercent rate for hadrons misidentified as muons, due to the extra shielding.

3.2.5.3 CMX

The CMX offered coverage in the forward region, $0.6 < |\eta| < 1.0$. It was composed of 2,208 drift tubes, arranged into conical sections at either end of the detector. The CMX sections were 8 layers deep, and were tilted slightly with respect to the beamline. A muon had to have $p_T \gtrsim 1.4$ GeV/c to reach the CMX. Scintillating tiles were present on the outside surfaces of the CMP and CMX. They were used to compensate for the slow drift time of the muon drift tubes, which was greater than the period between two bunch crossing. A signal in the scintillators allowed the muon to be matched to the bunch crossing that produced it. Reconstruction of muons in this analysis is important since the signal and various control samples include muons in the final states.

3.3 Trigger and Data Acquisition Systems

The trigger system is a key element of any measurement at hadron colliders. At the typical Tevatron instantaneous luminosity, approximately two millions inelastic collisions per second occurred, corresponding to more than one interaction per bunch-crossing on average. Since the readout of the entire detector needs typically about 2 ms on average, after the acquisition of one event, another approximately 5,000 interactions would remain unrecorded. The CDF II front-end electronics was designed as to cope with this constraint and ensure that lost events were likely to be uninteresting for analysis. The percentage of events that were rejected solely because the trigger is busy processing previous events is referred to as trigger *deadtime*. The average size of the information associated to each event from the $\mathcal{O}(10^6)$ total CDF II channels was 300 kbytes. Even in case of deadtime less readout of the detector, in order to record all events an approximate throughput and storage rate of 600 Gbyte/s would be needed, largely beyond the possibilities of the available technology at the detector construction time. The maximum storage rate available at time was approximately 250 kb/s.

However, since the cross-sections of most interesting processes are 10^3–10^{12} times smaller than the inelastic $p\bar{p}$ cross-section, the above limitations could be overcome with an online preselection of the events that are most likely to be used in the offline analyses. This was the task of the trigger system, which evaluated in real time the information provided by the detector and discarded the uninteresting events.

The CDF II data acquisition (DAQ) comprises a three-level trigger that selectively reduced the acquisition rate with virtually no deadtime, i. e., keeping each event in the trigger memory for a time sufficient to allow for a trigger decision without inhibiting acquisition of the following events. Each level received the event accepted from the previous level by exploiting detector information of increasing complexity and more time for processing, the trigger applied a logical "OR" of several programmable selection criteria to make its decision (Fig. 3.8). Prior to any trigger, the bunched

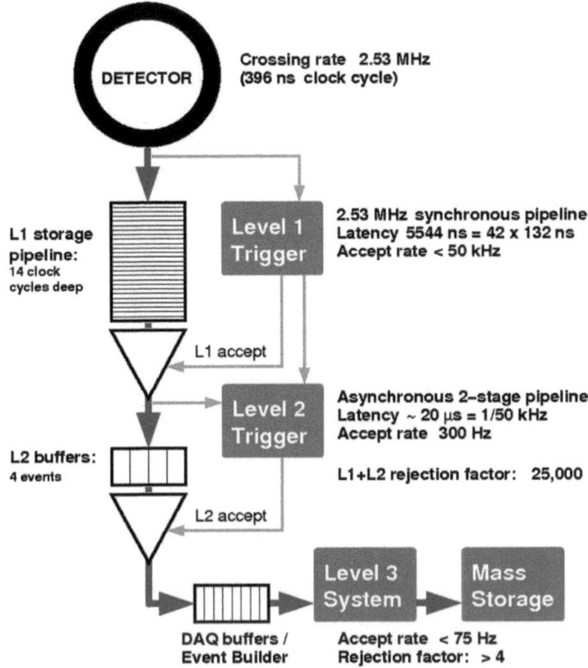

Fig. 3.8 Functional block diagram of the CDF II trigger and data acquisition systems

structure of the beams was exploited to reject cosmic-ray events by gating the front-end electronics of all subdetectors in correspondence of the bunch crossing.

The front-end of each sub-detector, located in electronic modules hosted in about 120 crates, had a 42-cells deep pipeline synchronized with the Tevatron clock-cycle set to 132 ns. The Tevatron clock picked up a timing marker from the synchrotron RF and forwarded this bunch-crossing signal to the trigger and to the front-end electronics. Since the inter-bunch time was 396 ns, the pipeline could collect data corresponding to a maximum of 14 bunch crossings. The pipeline depth defined the time available to *Level 1* (L1), $14 \times 396\,\text{ns} = 5.5\,\mu\text{s}$, for deciding to accept or reject an event, otherwise the buffer content was overwritten. An event accepted by the L1 was passed to the *Level 2* (L2) buffer, where the pipeline had 4 buffers, corresponding to $4 \times 5.5\,\mu\text{s} = 22\,\mu\text{s}$. If an event was accepted by the L1 and the L2 did not have a free buffer, deadtime was to incur. The L2 output rate was low enough to avoid further deadtime.

At L1, a synchronous system of custom-designed hardware processed a simplified subset of data in three parallel streams to reconstruct coarse information from the calorimeters (total energy and presence of single towers over threshold), the COT (two-dimensional tracks in the transverse plane), and the muon system (muon stubs). A decision stage combined the information from these low-resolution physics objects, called "primitives", into more sophisticated objects, e. g., track primitives were matched with muon stubs, or tower primitives, to form muon, electron, or jet objects, which then underwent some basic selections.[12]

At L2, an asynchronous system of special-purpose hardware processed the time-ordered events accepted by the L1. Additional information from the shower-maximum strip chambers in the central calorimeter and from the axial layers of the SVX II detector was combined with L1 primitives to produce L2 primitives. A crude energy-clustering was done in the calorimeters by merging the energies in adjacent towers to the energy of a seed tower above threshold. L1 track primitives matched with consistent shower-maximum clusters provided refined electron candidates whose azimuthal position was known with $2°$ accuracy. Information from the (r, ϕ) sides of the SVX II was combined with L1 tracks primitives to form two-dimensional tracks with offline-like resolution. Finally, an array of programmable processors made the trigger decision, while the L2 objects relative to the following event accepted at L1 were already being reconstructed.

The digitized output corresponding to the L2-accepted event reached *Level 3* (L3) via optical fibers and fragmented across all sub-detectors. It was collected by a custom hardware switch that arranged it in the proper order and transferred it to 300 commercial computers, organized in a modular, parallel structure of 16 subsystems. The ordered fragments were assembled in the *event record*, a block of data that univocally corresponded to a bunch-crossing and was ready for the analysis of the L3 software. The event reconstruction benefitted from full detector information and improved resolution with respect to the preceding trigger levels, including three-dimensional track reconstruction and tight matching between tracks and calorimeter or muon information. If an event satisfied the L3 requirements, the corresponding event record was transferred to mass storage at a maximum rate of 20 MB/s, approximately corresponding to 100 Hz. A fraction of the output was monitored in real time to search for detector malfunctions, derive calibrations constants, and graphically display events. The L3 decision was made after the full reconstruction of the event was completed and the integrity of its data was checked, a process that took a few milliseconds.

3.3.1 Dimuon Trigger

The main analysis sample used in this thesis was collected by the dimuon trigger. The dimuon trigger relied on a clear signature of two muons coming from $J/\psi \to \mu^+\mu^-$ decays. It used information from the COT and muon systems to identify muon candidates. This information was processed by two custom electronic processors, the extremely fast tracker (XFT) and the extrapolation unit (XTRP). Track primitives with $p_T \gtrsim 1.5$ GeV were identified by the XFT. The XFT coarsely determined p_T and ϕ_6 (the ϕ of the particle measured at a radius corresponding to COT superlayer 6) of

---

[12] A jet is a flow of observable secondary particles produced in a spatially collimated form, as a consequence of the hadronization of partons produced in the hard collision.

the track using the four axial superlayers of the COT. The XFT determined track p_T, ϕ, and charge with an uncertainty of $\sigma_{p_T}/p_T^2 \approx 1.7\%$, $\sigma_\phi \approx 5$ mrad, about ten times worse than the offline uncertainty. The XFT logic proceeded in two steps, segment finding and segment linking. During segment finding, wire hits were classified as "prompt" (drift time $<$ 66 ns) or "delayed" (drift time between 66 and 220 ns). Adjacent COT cells were grouped together four at a time and their hit information was compared to a set of predefined patterns to find track segments. Track segments were then grouped together in sets of four (each from a different superlayer) to give a coarse estimate of track parameters. The predefined patterns ensured that all found segments originated close to the beamline and had a high enough p_T. The XTRP received track information from the XFT and geometrically extrapolated the tracks to the calorimeters and muon detectors seeking for matches in energy deposition or muon stubs. Muon and dimuon primitives were derived from hits in the muon chambers.

The dimuon trigger was a combination of two triggers: CMU–CMU, where both muons were identified in the centralmost muon chamber, and CMU–CMX, where one muon was found in CMU and one in CMX. We only describe the CMU–CMU trigger here, and comment on the differences of the CMU–CMX.

A stack was a set of four drift cells stacked on top of each other. The CMU had 288 stacks in each of the East and West sides of the detector. A Level 1 stub was a track segment in a stack such that cells 1 and 3 or cells 2 and 4 have hits separated by no more than 396 ns. A tower was a set of two neighboring stacks. A tower had fired when a Level 1 stub was identified in one or both stacks. A muon tower was a tower that fired, matched with an XFT track. In order to keep the Level 1 decision-time short enough to remain synchronous, only information about which towers have fired was used in triggering, rather than detailed hit positions and direction. If a tower was found within a 3σ window in ϕ of the tracker extrapolation (wide enough to account for multiple Coulomb scattering), it was considered a muon tower. The CMU–CMU trigger required that at least two muon towers be found such that they were either on opposite sides of the detector or were separated by at least two other towers. The CMU–CMX trigger used a similar algorithm. In the CMU–CMX case, only XFT tracks with $p_T > 2.2$ GeV/c were used to match to the CMX tower, as the extra material that muons passed through to reach the CMX further constrained the momentum requirements on the muon, and no azimuthal separation was required because the muons were by construction in different subdetector volumes. The dimuon trigger underwent steady improvements over the course of Run II. While the core logic outlined above was unchanged, some parameters were changed often to improve the trigger including requirements on the p_T of the XFT tracks, the difference in ϕ between the two muons, and their transverse mass M_T. In addition, some of the triggers were prescaled, which means that only one out of N events that met the trigger selection criteria were accepted. This reduced the load on the DAQ system at higher luminosity. The various combinations of these requirements resulted in slightly different trigger requirements across the data sample.

3.3.2 Reconstruction of Physics Objects

The raw outputs of several CDF subdetectors were combined in order to reconstruct
the high-level physics objects such as muons and tracks, that are used in the analysis.

3.3.2.1 Tracking

Within an uniform axial magnetic field, the trajectory of a charged particle produced
with non-zero initial velocity in the bending plane of the magnet is described by an
helix, which can be uniquely parameterized by the following set of equations:

$$\begin{cases} x = R\sin(2Cs + \phi_0) - (R + d_0)\sin\phi_0, \\ y = -R\cos(2Cs + \phi_0) + (R + d_0)\cos\phi_0, \\ z = z_0 + s\lambda. \end{cases}$$

Given the projected length along the track s, the corresponding (x, y, z) coordinates
of the trajectory are functions of five parameters (Fig. 3.9):

C signed half-curvature of the helix, defined as $C = q/2R$, where R is the radius
    of the helix and q is the charge of the particle. This is directly related to the
    transverse momentum, $p_T = cB/(2|C|)$, where c is the speed of light and B is
    the magnetic field;

ϕ_0 ϕ angle of the particle trajectory at the point of closest approach to the z axis;

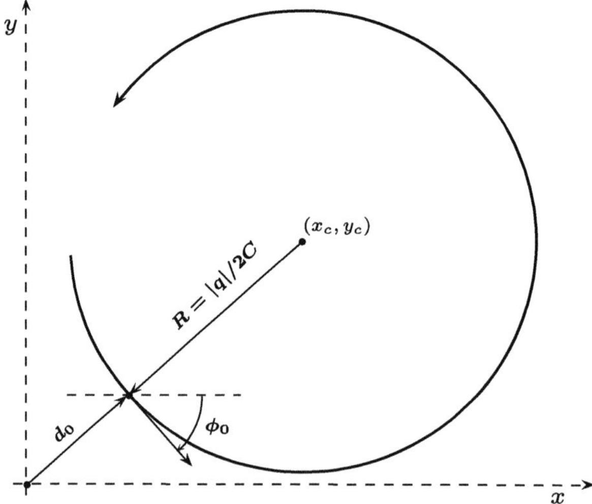

Fig. 3.9 Schematic view of a positively charged track in the plane transverse to an axial magnetic
field $\boldsymbol{B} = (0, 0, -B)$

d_0 signed impact parameter, i.e., the radial distance of closest approach to the z axis, defined as $d_0 = q(\sqrt{x_c^2 + y_c^2} - R)$, where (x_c, y_c) are the coordinates of the center;

λ the helix pitch, i.e. $\cot\theta$, where θ is the polar angle of the helix at the point of its closest approach to the z axis. This is directly related to the longitudinal component of the momentum $p_z = p_T \cot\theta$;

z_0 the z coordinate of the point of closest approach.

The reconstruction of a charged-particle trajectory consisted in determining the helix parameters trough a fit of the hits reconstructed in the tracking sub-detectors with two basic steps: clustering multiple hits likely to be originated from the same particle, and using pattern-recognition algorithms to join the hits along the whole track arc to find the final trajectory. Non-uniformities of the magnetic field and multiple-scattering effects in the material were accounted for in the fit. CDF employed several algorithms for track reconstruction. The main algorithm, used in this measurement, is the *Outside-In* (OI). Tracks are first reconstructed in the COT and then extrapolated inward to the silicon tracker. The greater radial distance of the COT with respect to the silicon tracker resulted in a lower track density and consequent fewer accidental combination of hits in the initial steps of track reconstruction. A concise overview of all the algorithms used at CDF is given in Refs. [13–15], in the following we briefly summarize the OI working principle and its typical perfomances.

In the first step of pattern recognition, cells in the axial super-layers were searched for sets of four or more hits that were fit to a straight line. Once these "segments" of hits were found, two approaches were followed to reconstruct a track. Segments consistent with lying tangent to a common circular path were linked together. The helical track, when projected onto the (r, ϕ) plane, is a circle. An alternative approach was to constrain its circular fit to the beam-line. Once a circular path was found in the (r, ϕ) plane, segments and hits in the stereo super-layers were added depending on their proximity to the circular fit. This resulted in a three-dimensional track fit. Once a track was reconstructed in the COT, it was extrapolated inward to the silicon detectors. Based on the estimated uncertainties on the track parameters, a three-dimensional "road" was formed around the extrapolated track. Starting from the outermost layer inwards, silicon hits found inside the road were added to the track. As hits were added, the searched road narrowed, according to the updated knowledge of track parameters and their covariance matrix. Reducing the width of the road reduced the chance of adding an unrelated hit to the track and the computation time for reconstruction. In the first pass of this algorithm, only axial hits were considered. In a second pass, hits with stereo information were added. Finally, the track combination associated to the highest number of hits and having the best fit quality χ^2/ndf was kept.

The COT efficiency for particles with p_T larger (smaller) than 1 GeV/c was typically 98–99 % (95 %) depending on whether the particle was isolated. The typical resolutions on track parameters were $\sigma_{p_T}/p_T^2 \approx 0.15\,\%$ (p_T in GeV/c), $\sigma_{\phi_0} \approx 0.035°$, $\sigma_{d_0} \approx 250\,\mu\mathrm{m}$, $\sigma_\theta \approx 0.17°$ and $\sigma_{z_0} \approx 0.3\,\mathrm{cm}$ for tracks reconstructed with no use of silicon information, and not constrained to originate from the beam. The silicon information improved the impact parameter resolution which, depending on the number (and radial distance) of the silicon hits, could reach $\sigma_{d_0} \approx 15\,\mu$m (not including the transverse beam size). This value, combined with the typical $30\,\mu$m transverse beam size, was sufficiently small with respect to the typical transverse decay-lengths of heavy flavors (a few hundred microns) to allow separation of their decay-vertices from the production vertices. The silicon tracker improves also the stereo resolutions to $\sigma_\theta \approx 0.06°$ and $\sigma_{z_0} \approx 70\,\mu$m, while transverse momentum resolution can be further improved to about $\sigma_{p_T}/p_T^2 \approx 0.07\,\%$ with p_T in GeV$/c$.

3.3.2.2 Muons

Muon identification is crucial for the analysis described in this thesis; the dimuon (J/ψ) trigger was used to select events enriched in $J/\psi \to \mu\mu$ decays. As minimally ionizing particles, muons do not emit significant bremsstrahlung radiation in the calorimeters, and very rarely interact with nuclei. They passed through the whole detector depositing little energy. Hence, muons with enough momentum reached the muon detectors. The design of the detector was such that other particles (excluding neutrinos) were likely to be absorbed by the material between the beam and the first of the muon detectors. A particle entering the muon chambers left a track-segment which was registered as a *muon stub*. The extrapolation of a track reconstructed in the COT matched geometrically the orientation of the muon stub. A muon candidate was formed. Occasionally, though, an energetic hadronic particle could decay towards the back of the hadronic calorimeter, and its decay products could "punch through" to the muon chambers. These particles produce experimental signatures similar to those from genuine muons and were called "false muons".

3.4 Monte Carlo Simulation

Estimation of the fraction of events of a certain type that escape the detector acceptance, or detailed studies of the average expected response of the detector to the passage of particles are common needs in many analyses. Usually, complex detector geometries and the numerous effects that need to be accounted for in predicting their response make the analytical derivation of the relevant distributions impractical or impossible. Monte Carlo techniques provide a useful and widely-used complement. We provide here a short overview of the standard CDF II simulation. Further details can be found in Ref. [16].

In the standard CDF II simulation, the detector geometry and material were modeled using version 3 of the GEANT package [17] tuned to test-beam and collision data. GEANT received in input the positions, the four-momenta, and the identities of all final-state particles that have long enough lifetimes to exit the beam pipe. It simulated their passage in the detector, modeling their interactions (*bremsstrahlung*, multiple scattering, nuclear interactions, photon conversions, etc.) and the consequent generation of signals on a single channel basis. Specific custom packages substituted GEANT for some subdetectors. The calorimeter response was simulated with GFLASH, a faster parametric shower-simulator [18] tuned for single-particle response and shower-shape using test-beam data (8–230 GeV electrons and charged pions) and collision data (0.5–40 GeV/c single isolated tracks); the drift-time within the COT was simulated using the GARFIELD standard package [19] further tuned on data; the charge-deposition model in the silicon used a parametric model, tuned on data, which accounted for restricted Landau distributions, production of δ-rays, capacitive charge-sharing between neighboring strips, and noise [20].[13] Furthermore, the actual trigger logic was simulated. The output of the simulated data mimicked the structure of experimental data, allowing their analysis with the same reconstruction programs.

The detector and trigger configuration underwent variations during data-taking. Minor variations occurred between runs, while larger variations occurred after major hardware improvements, or Tevatron shut-down periods. For a more detailed modeling of the actual experimental conditions, the simulation had been interfaced with the offline database that reported, on a run-by-run basis, all known changes in configuration (position and slope of the beam line, relative mis-alignments between subdetectors, trigger-selection used) and local or temporary inefficiencies in the silicon tracker (active coverage, noisy channels, etc.). This permitted simulating the detailed configuration of any set of real runs to use it, after proper luminosity reweighing, for modeling the realistic detector response in any given subset of data. Although we chose a strictly data-driven approach for this analysis, limited use of simulation was made to determine detector acceptances, optimize the selection requirements, and develop the algorithms used for identification of the bottom strange meson flavor at production. Montecarlo B_s^0 signal decays used in this thesis are indeed simulated according to the phase space from EvtGen available to the decay averaging over the spin states of the decay daughters. This ensures flat distributions of the angles between final state particles. B mesons are generated within a pseudorapidity of 1.3 and above a p_T threshold of 4 GeV/c. The transverse momentum spectrum is tuned according to data, exploiting informations from inclusive J/ψ cross-section measurements. The masses and lifetimes of the particles are taken from PDG of 2010.

References

1. S. D. Holmes (ed.), *Tevatron Run II Handbook* (FERMILAB-TM-2484, 1998)
2. S. Holmes, R.S. Moore, V. Shiltsev, Overview of the Tevatron collider complex: goals, operations and performance (2011), arXiv:1106.0909 (physics.acc-ph)
3. D. Mohl et al., Physics and technique of Stochastic cooling. Phys. Rept. **58**, 73 (1980)
4. S. Nagaitsev et al., Experimental demonstration of relativistic electron cooling. Phys. Rev. Lett. **96**, 044801 (2006)

---

[13] The δ-rays are knock-on electrons emitted from atoms when the passage of charged particles through matter results in transmitted energies of more than a few keV in a single collision.

5. R. Blair et al. (CDF Collaboration), *The CDF II Detector: Technical Design Report* (FERMILAB-DESIGN-1996-01, 1996)
6. C.S. Hill (On behalf of the CDF Collaboration), Operational experience and performance of the CDF II silicon detector. Nucl. Instrum. Meth. **A530**(1) (2004)
7. A. Sill (CDF Collaboration), CDF Run II silicon tracking projects. Nucl. Instrum. Meth. **A447**(1) (2000)
8. A.A. Affolder et al. (CDF Collaboration), Intermediate silicon layers detector for the CDF experiment. Nucl. Instrum. Meth. **A453**, 84 (2000)
9. A.A. Affolder et al. (CDF Collaboration), CDF central outer tracker. Nucl. Instrum. Meth. **A526**, 249 (2004)
10. D. Acosta et al. (CDF Collaboration), A time-of-flight detector in CDF II. Nucl. Instrum. Meth. **A518**, 605 (2004)
11. G. Ascoli et al., CDF Central Muon Detector. Nucl. Instrum. Meth. **A268**, 33 (1988)
12. C. Ginsburg (on behalf of the CDF Collaboration), CDF Run 2 muon system. Eur. Phys. J. **C33**, S1002 (2004)
13. P. Gatti, *Performance of the new tracking system at CDF II*, Ph.D. thesis, University of Pisa, available as FERMILAB-THESIS-2001-23, 2001
14. S. Menzemer, *Track reconstruction in the silicon vertex detector of the CDF II experiment*, Ph.D. thesis, Karlsruhe University, available as FERMILAB-THESIS-2003-52, IEKP-KA-2003-04, 2003
15. C.P. Hays et al., Inside-out tracking at CDF. Nucl. Instrum. Meth. **A538**, 249 (2005)
16. E. Gerchtein, M. Paulini, CDF detector simulation framework and performance, eConf C0303241, TUMT005 (2003), arXiv:physics/0306031 (physics)
17. R. Brun, R. Hagelberg, M. Hansroul, J. Lassalle, GEANT: Simulation program for particle physics experiment. User guide and reference manual (1978)
18. P.A. Movilla Fernandez (CDF Collaboration), Performance of the CDF calorimeter simulation in Tevatron Run II, in *Proceedings of AIP Conference* 867, 487 (2006), arXiv:physics/0608081
19. R. Veenhof, GARFIELD, recent developments. Nucl. Instrum. Meth. **A419**, 726 (1998)
20. M. Gold, Description of the parameterized charge deposition model. CDF Note **5871** (2002)

Chapter 4
Analysis Selection

*With this chapter the actual description of the analysis begins. It describes the
selection and reconstruction of the experimental data samples used in the measurement.*

4.1 The Decays

The data used in this measurement were collected using the CDF II detector between
February 2002 and September 2011. The reconstructed decay modes are (charge-
conjugate modes implied)

- $B_s^0 \to J/\psi\phi$ followed by the $J/\psi \to \mu^+\mu^-$ and $\phi \to K^+K^-$ decays,
  for the measurement of $\beta_s^{J/\psi\phi}$.
- $B^+ \to J/\psi K^+$ followed by the $J/\psi \to \mu^+\mu^-$ decays
  for the validation and calibration of the algorithms used in the identification of
  the initial bottom meson flavor that use information from the other b hadron in the
  event (Sect. 5.3.2).
- $B_s^0 \to D_s^-\pi^+$ followed by

  - the $D_s^- \to \phi\pi^-$ and $\phi \to K^+K^-$ decays;
  - the $D_s^- \to K^*(892)^0 K^-$ and $K^*(892)^0 \to K^+\pi^-$ decays;
  - the $D_s^- \to \pi^+\pi^+\pi^-$ decays; and

- $B_s^0 \to D_s^-\pi^+\pi^+\pi^-$ followed by the $D_s^- \to \phi\pi^-$ and $\phi \to K^+K^-$ decays,
  for the validation of the algorithms used in the identification of bottom meson
  flavor that use information from the signal bottom meson (Sect. 5.3.3).

The reconstruction of $B^+ \to J/\psi K^+$, $B_s^0 \to D_s^-\pi^+$, and $B_s^0 \to D_s^-\pi^+\pi^+\pi^-$
calibration samples is outlined in Chap. 5. In the following we only detail the selection
of the $B_s^0 \to J/\psi\phi$ sample, since this is the signal used in the measurement.

© Springer International Publishing Switzerland 2015

S. Leo, *CP Violation in $B_s^0 \to J/\psi\phi$ Decays*, Springer Theses,

DOI 10.1007/978-3-319-07929-5_4

4.2 Trigger and Dataset

The data collected by the di-muon triggers throughout the whole Run II sample, corresponding to an integrated luminosity of $\mathcal{L} \approx 9.6\,\mathrm{fb}^{-1}$, are selected based on trigger requirements at L1, with L2 and L3 playing a minor role in the trigger decision. L2 is used to tighten any existing requirements of L1, e.g., on the transverse momentum, and L3 uses more precise determination of several event variables, such as the transverse momentum of charged particles, better track-stub matching information, dimuon mass, and so forth (Sect. 3.3.1). The integrated luminosity refers to the value corresponding to the sample resulting from a quality selection that discards all data collected when a detector subsystem relevant for this analysis was malfunctioning or off.

4.3 Event Sample Selection and Signal Reconstruction

Collected data are subjected to an offline reconstruction. This includes fitting tracks, combining them into vertices, reconstructing decay chains, and constructing the higher-level physics quantities in a convenient format for offline analysis.

4.3.1 Preselection

Simple requirements on directly-observed quantities allow a first baseline selection to remove events that most likely do not contain a real B_s^0 decay. The goal is to increase the purity, while keeping the efficiency for signal nearly unchanged. An important part of the preselection requirements are selection criteria on track and vertex quality. The fraction of false or mismeasured tracks is reduced by requiring a minimum number of hits in the drift chamber and silicon detectors. Such standard selection requirements were studied in dedicated measurements at the beginning of Run II to yield optimal track-finding efficiency at sufficient purity. Additional kinematic requirements complete the set of preselection criteria. The main preselection criteria are as follows:

- Track quality

  - >9 axial and >9 stereo COT hits and >3 silicon hits for each track;
  - $p_T > 0.4\,\mathrm{GeV}/c$.

- Vertex quality

  - $\chi^2_{r\phi} < 50$, where $\chi^2_{r\phi}$ is the χ^2 of the two-dimensional $r\phi$ vertex fit.

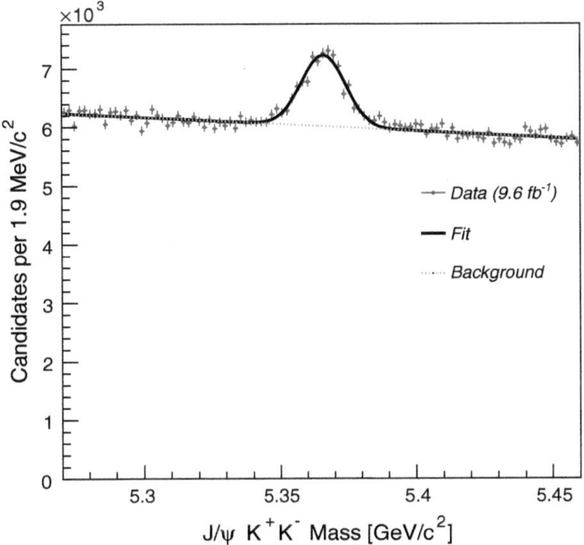

Fig. 4.1 Candidate mass distribution after event preselection

- Kinematic requirements

  - $5.1 < M(J/\psi KK) < 5.6\,\mathrm{GeV}/c^2$;
  - $p_T(\phi) > 1\,\mathrm{GeV}/c$;
  - $3.014 < M(\mu\mu) < 3.174\,\mathrm{GeV}/c^2$;
  - $1.009 < M(KK) < 1.028\,\mathrm{GeV}/c^2$;
  - $p_T(B_s^0) > 4\,\mathrm{GeV}/c$.

We use the known J/ψ meson mass as constraint on the di-muon mass in fitting the B_s^0 meson decay. The resulting $B_s^0 \to J/\psi K^+K^-$ mass is shown in Fig. 4.1. A narrow structure is visible containing about 13,000 events and overlapping a significant background, corresponding to a signal-to-background ratio (S/B) of about 7.3 % integrated within 3σ from the B_s^0 mass. To further improve the $\beta_s^{J/\psi\phi}$ measurement, a neural network classifier is used to optimize S/B.

4.3.2 Neural Network Selection

We use an artificial neural network (NN) selection to combine the information from a set of discriminant distributions into a single output variable that determines the probability for a candidate to be signal-like or background-like on a conventional scale of $+1.0$ (signal) to -1.0 (background). Through a machine-learning process, the kinematic variables and particle identification informations are assigned weights associated to their estimated discriminating power. Correlations between variables

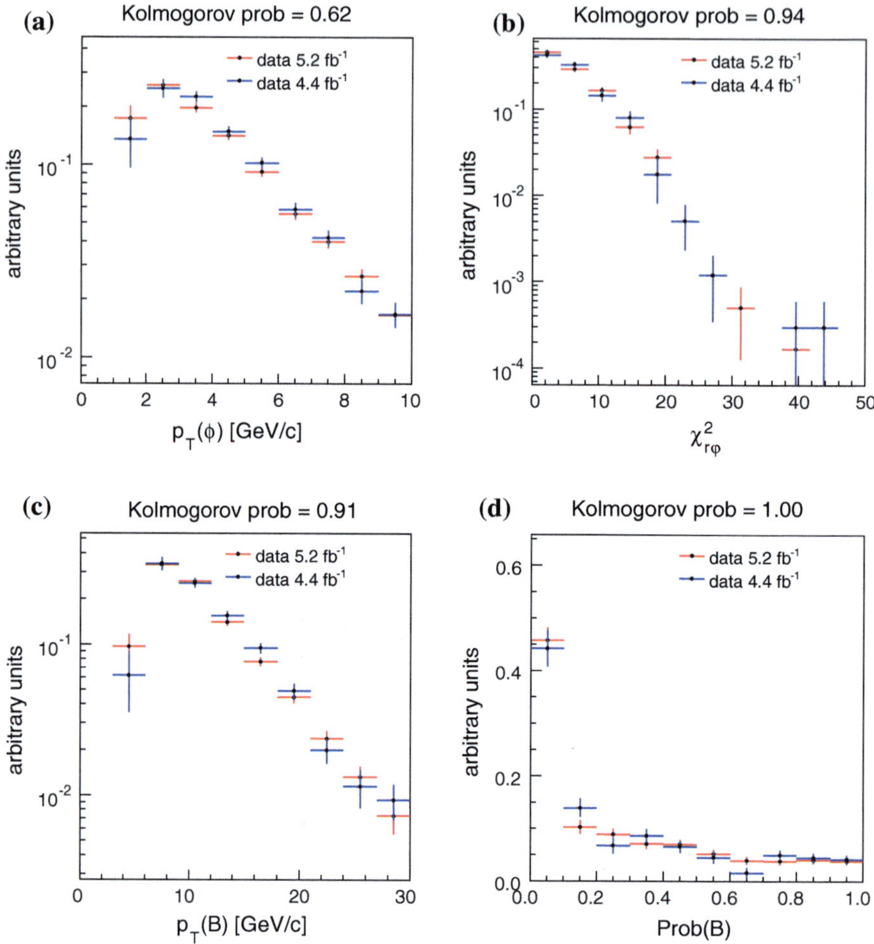

Fig. 4.2 Comparison between the sideband-subtracted distributions of NN input variables in the first 5.2 fb^{-1} (*red*) and the last 4.4 fb^{-1} (*blue*) of data showing $p_T(\phi)$ (**a**) and $\chi^2_{r\varphi}$ (**b**), $p_T(B)$ (**c**), χ^2 probability of the B_s^0 vertex fit (**d**) (color figure online)

are exploited by the neural network. If properly trained and protected against biases, a NN achieves an improvement in efficiency on signal and in background rejection over any combination of conventional one-dimensional requirements on directly observed quantities.

In this work we use the same NN classifier that was trained and optimized in detail in previous versions of the analysis, based on a subset of the present data [1, 2]. Because the relevant data distributions are proved to be reasonably uniform in time, we expect the current performance of the NN discriminator to be comparable with the performance previously obtained.

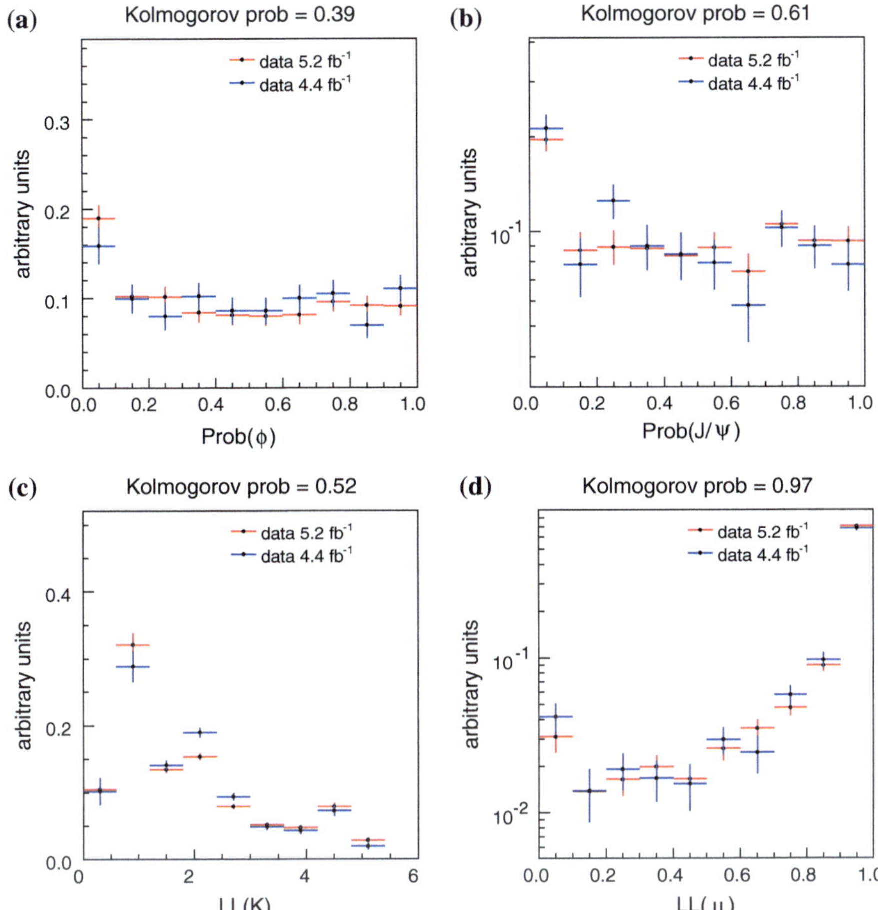

Fig. 4.3 Comparison between the sideband-subtracted distributions of NN input variables in the first 5.2 fb^{-1} (*red*) and the last 4.4 fb^{-1} (*blue*) of data showing χ^2 probability of the ϕ vertex fit (**a**) and χ^2 probability of the J/ψ vertex fit (**b**), $LL_K(K)$ (**c**), $LL_\mu(\mu)$ (**d**)

The artificial neural network is constructed using the `NeuroBayes` package [3]. To assign weights to the input variables, the neural network is fed with events representative of signal and events representative of background, which are used to identify patterns to discriminate background-like and signal-like event features. For training the NN to identify signal, we use simulated events reproducing the trigger and detector conditions present in experimental data, while the background sample comprise 0.3 million data events from the B_s^0 meson mass sidebands, (5.291 < $M_{J/\psi K^+ K^-}$ < 5.315 GeV/c^2) ∪ (5.417 < $M_{J/\psi K^+ K^-}$ < 5.442 GeV/c^2).

Some of the most important variables input to the neural network are listed in Table 4.1. We use the χ^2 of the three vertex fits in three dimensions for each par-

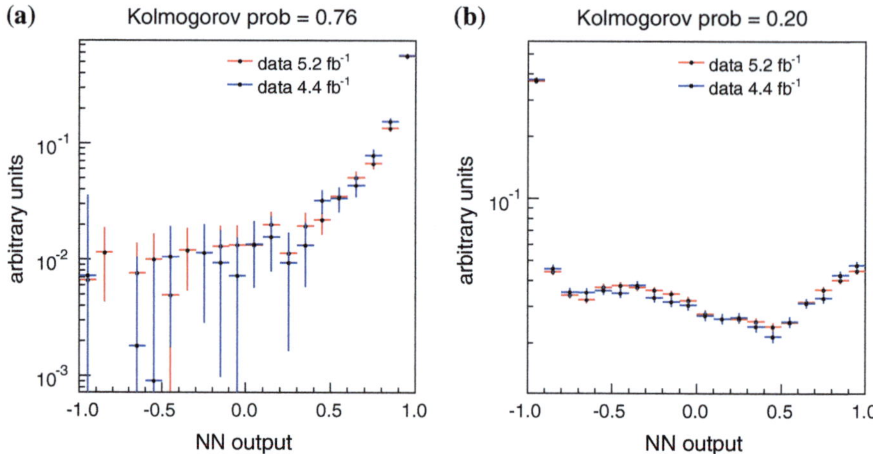

Fig. 4.4 Comparison between the NN output distributions in the first 5.2 fb^{-1} of data (*red*) and the remaining 4.4 fb^{-1} of data (*blue*). The NN output variable for (sideband-subtracted) signal events only (**a**) and for sidebands events only (**b**)

Table 4.1 Variables used for NN training, and associated importance ranking

Variables			Ranking
Vertex fit probability		$\chi^2(\phi)$	9
		$\chi^2(B_s^0)$	8
		$\chi^2(J/\psi)$	10
		$\chi^2_{r\phi}(B_s^0)$	5
Transverse momentum		$p_T(\phi)$	1
		$p_T(B_s^0)$	6
Particle ID		CLL(K_1)	3
		CLL(K_2)	2
		$LL_\mu(\mu_1)$	7
		$LL_\mu(\mu_2)$	4

ticle of interest, the B_s^0, J/ψ, and ϕ mesons. In addition, we use $\chi^2_{r\phi}$, the measure of goodness of the transverse-plane fit of the B_s^0 vertex. We include the reconstructed masses of the vector mesons J/ψ and ϕ, and the transverse momenta of the three decaying particles. We use the combined log likelihood ratio (CLL) for K/π separation [4]. Finally, we use the maximum and the minimum of the two muon likelihood values LL_μ [5]. The muon and kaon likelihoods are quantities used for particle identification. The kaon likelihood [4] is a combined discriminant constructed from the kaon specific-ionization energy-loss, dE/dx, and its time-of-flight information, while the muon likelihood exploits track-stub matching quantities along with energy deposition information in electromagnetic and hadronic calorimeters. Sideband events are subtracted from signal events in data, which populate the

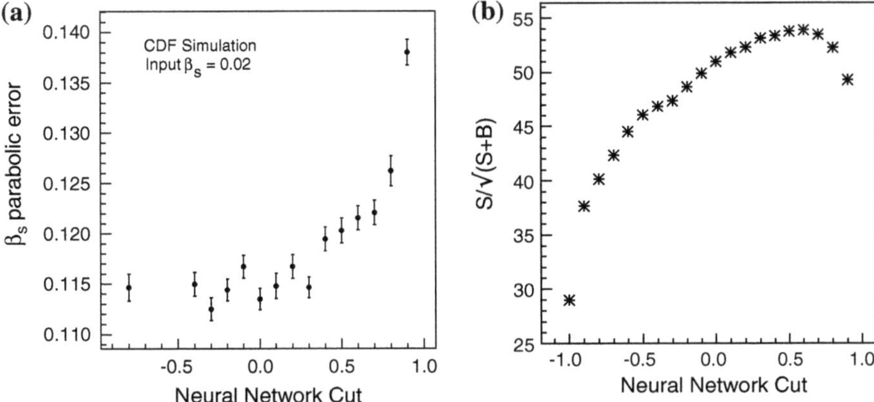

Fig. 4.5 Most probable value of the statistical uncertainty on $\beta_s^{J/\psi\phi}$ as a function of the NN threshold for a true value of $\beta_s^{J/\psi\phi} = 0.02$ in pseudoexperiments (**a**) Figure of merit $S/\sqrt{S+B}$ as function of NN threshold (**b**)

region $5.34 < M_{J/\psi K^+ K^-} < 5.39\,\mathrm{GeV}/c^2$, corresponding roughly to a $\pm 2.5\sigma$ range around the signal peak, to obtain a statistically pure sample of signal events. We compare the sideband-subtracted distributions of all network input variables in the first 5.2 fb^{-1} of data and the second 4.4 fb^{-1} (Figs. 4.2, 4.3). We expect this comparison to be sensitive to effects such as changes in the trigger composition of our sample, since kinematic thresholds may differ in each trigger selection and may have evolved in time; and other known or unknown changes in operation conditions. Figs. 4.2, 4.3, and 4.4 show the results of the comparison. No relevant discrepancy is observed as indicated by Kolmogorov-Smirnov tests [6]. Even in case of moderate mismodeling, it is unlikely that any selection bias would be induced in the final results of the measurement, which would only be affected by a suboptimal selection.

The input variables are ranked by discriminating power in the NN, resulting in ϕ transverse momentum and kaon likelihoods as being the most powerful discriminators (Table 4.1).

The NN output threshold is chosen so as to minimize the expected average variance on the estimate of β_s as determined in large samples of statistical trials generated with different choices of true values for the phase and the width difference. The $\beta_s^{J/\psi\phi}$ statistical uncertainty distributions corresponding to each NN output threshold value are fit with an appropriate function to determine the most probable value of the variance on the mixing phase estimate. The results of these fits are studied as function of the NN threshold. Figure 4.5a shows the studies conducted using the example true value $\beta_s^{J/\psi\phi} = 0.02$. The resulting optimal NN threshold value is approximately 0.2, largely independent of the true $\beta_s^{J/\psi\phi}$ value. This corresponds to a significantly more inclusive selection criteria than would be selected by a commonly used optimization procedure that maximize the figure of merit $S/\sqrt{S+B}$ (where S and B are the number of signal and background events, respectively) as represented in Fig. 4.5b.

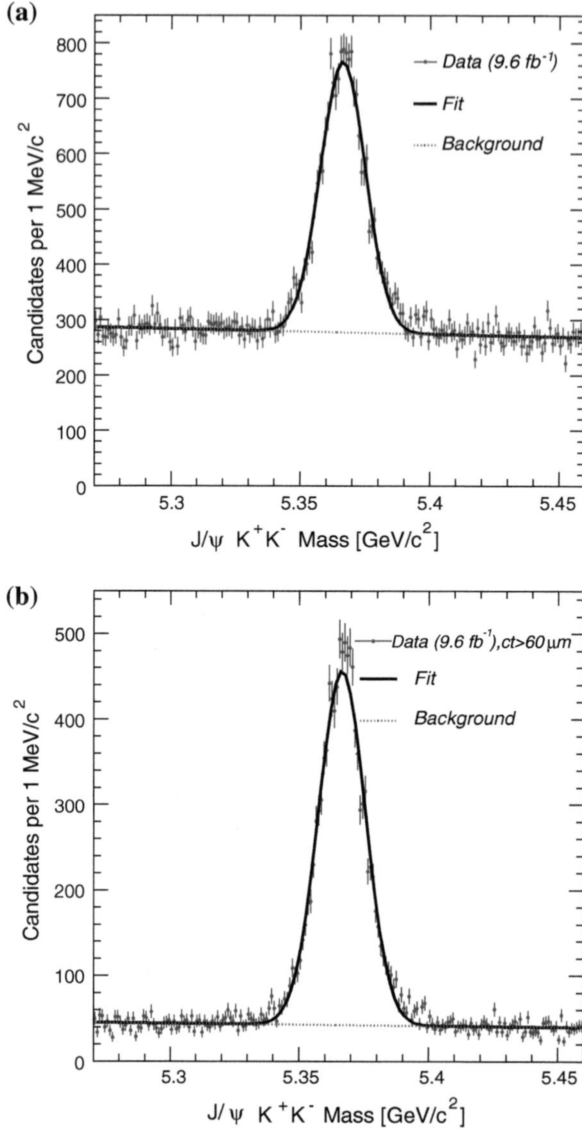

Fig. 4.6 $J/\psi K^+ K^-$ mass distribution for the full Run II data sample (**a**) and same distribution after an additional $ct > 60\,\mu$m requirement on the decay length (**b**)

4.3.3 The Final Sample

The $J/\psi K^+ K^-$ mass distribution of the resulting sample is shown in Fig. 4.6a. The sample composition can be coarsely described in terms of three classes of events.

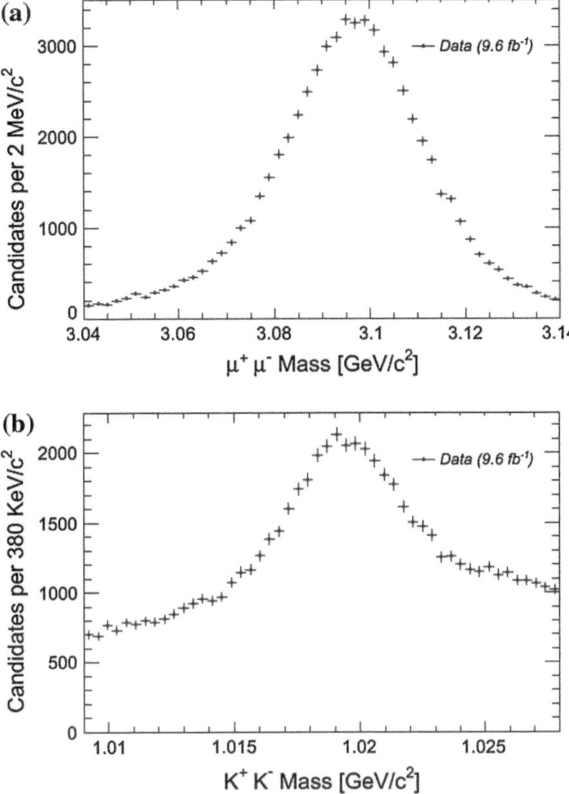

Fig. 4.7 Mass distribution of reconstructed J/ψ (**a**) and ϕ (**b**) decays for the full Run II data sample

Signal: a prominent peak of approximately 11,000 signal B_s^0 events is visible over a smooth background. The width of the signal is about 9 MeV/c^2, dominated by experimental resolution. The signal purity is $S/B \approx 76\,\%$. Another source of signal may include non-resonant $J/\psi K^+ K^-$ decays or $J/\psi K^+ K^-$ events in which the kaon pair originates from a $f_0(980)$ decay. These contributions cluster at the same mass where the $J/\psi\phi$ decay peaks. In the analysis we include their contribution in the full decay rate (Sect. 2.4.1).

Combinatorial background: these events are mainly real J/ψ decays combined with pairs of random charged particles that satisfy accidentally the selection requirements. Indeed, the dimuon mass spectrum in Fig. 4.7a as reconstructed in data shows a narrow enhancement at the known J/ψ mass over a sparsely populated background, indicating that the dimuon sample contains genuine $J/\psi \rightarrow \mu^+\mu^-$ decays with high purity. Combination of real J/ψ decays with random tracks have a continuous, approximately flat mass distribution in the final B_s^0 mass spectrum. This is the main source of background in the analysis and it is distributed predominantly at short decay lengths, as shown in Fig. 4.6b, where the mass spectrum of events restricted to the decays with $ct(B_s^0) > 60\,\mu$m is shown for illustration purposes.

Physics background: given the limited particle identification capabilities, the sample presumably contains a $\mathcal{O}(\%)$ fraction of $B^0 \rightarrow J/\psi(\rightarrow \mu^+\mu^-)K^*(\rightarrow K^+\pi^-)$ decays misreconstructed as signal (B^0 cross-feed). This effect is visible as a small mass mismodelling at $M(J/\psi K^+ K^-) \approx 5.4\,\mathrm{GeV}/c^2$. It is studied in detail and a systematic uncertainty is associated (Sect. 7.1 and Appendix A).

References

1. L.B. Oakes, Measurement of the CP Violating Phase β_s in $B_s^0 \rightarrow J/\psi\phi$ Decays (2010)
2. E. Pueschel, Measurement of the CP Violating Phase $\sin(2\beta_s)$ using $B_s^0 \rightarrow J/\psi\phi$ Decays at CDF (2010)
3. M. Feindt, A Neural Bayesian Estimator for Conditional Probability Densities, arXiv:physics/0402093
4. M. Milnik, Measurement of the lifetime difference and cp-violating phase in $B_s \rightarrow J/\psi\phi$ decays. Ph.D. thesis, FERMILAB-THESIS-2007-38 (2007)
5. G. Giurgiu, B flavor tagging calibration and search for B(s) oscillations in semileptonic decays with the CDF detector at fermilab, Ph.D. thesis, Carnegie Mellon University, FERMILAB-THESIS-2005-41 (2005)
6. F. James, Statistical Method in Experimental Physics (World Scientific Publishing Co., Pte. Ltd., 2006)

Chapter 5
Preparation of Tools

This chapter reports the preparation and calibration of tools needed in the fit to the time evolution, including the determination of the detector angular acceptance and the calibration of the performance of flavor tagging algorithms.

5.1 Angular Acceptance

Since the final state $J/\psi\phi$ is an admixture of CP-eigenstates, the CP-even and -odd components are separated using differences in the angular distributions of final-state particles. Three angles are needed to specify the angular distributions. In the transversity basis (Sect. 2.4) we chose these to be the polar angle Θ and the azimuthal angle Φ of the positive muon direction in the J/ψ rest frame, and the transversity angle Ψ of the positive kaon direction in the ϕ rest frame (Sect. 2.4). The four-momenta of all signal final-state particles in the laboratory frame are reconstructed by the tracking system. From these, the four-momenta of the B_s^0, the J/ψ, and the ϕ mesons are determined. The Lorentz boosts are derived and all needed four-momenta are determined in the rest frame of any reconstructed particle to extract the needed angles.

The detector acceptance and the offline analysis selection affect the observed distribution of the angular variables through an acceptance $\varepsilon(\Psi, \Theta, \Phi)$. We parametrize the acceptance using a set of basis functions in three dimensions in the ranges $0 < \Psi < \pi, 0 < \Theta < \pi$, and $0 < \Phi < 2\pi$:

$$\varepsilon(\Theta, \Phi, \Psi) = a^k{}_{lm} P_k(\cos \Psi) Y_{lm}(\Theta, \Phi). \tag{5.1}$$

The detector acceptance in the above expression is parametrized through the coefficients $a^k{}_{lm}$. These efficiency coefficients are fit in a sample of 100 million signal-only $B_s^0 \rightarrow J/\psi\phi$ decays, simulated according to a phase-space decay-model. Since all spins of the particles in the final state are averaged, the angular

© Springer International Publishing Switzerland 2015
S. Leo, *CP Violation in $B_s^0 \rightarrow J/\psi\phi$ Decays*, Springer Theses,
DOI 10.1007/978-3-319-07929-5_5

variables assume constant distributions. The simulated events are subjected to the same detector acceptance, preselection, and NN selection used in data, which bias these distributions. Hence, the resulting angular distributions of simulated events are used as a model of the angular acceptance.

5.1.1 Simulation Reweighting

The quality of the determination of the transversity-angle sculpting depends on the agreement between data and simulation in those variables that affect the angular decay features of J/ψ and ϕ mesons. The agreement between data and simulation is investigated in several variables using sideband-subtracted data. A known mismodeling [1, 2] in the $p_T(B)$ spectrum between simulation and data is shown in Fig. 5.1a as an example. Since trigger prescales modify the trigger composition of the sample, and different trigger selections may have different p_T thresholds, a prescale-dependent reweighing is also applied. In the following, we compare sideband-subtracted data distributions with simulation distributions reweighed to match data as appropriate. The reweighting involves three steps: the first step corrects for the admixture of different trigger selections present in data; the second step imposes the agreement in the $p_T(B_s^0)$ spectra, and the purpose of the third step is to account for the combined effect of both the previous effects to eliminate any residual dependencies.

Trigger selection mixture: the candidates are divided into two groups, according to whether they are compatible with triggering a CMU–CMU or a CMU–CMX muon pair. Each of the two classes is then split in three classes defined as events with both muons having $p_T > 3$ GeV/c, with muons having $p_T > 2$ GeV/c and at least one muon has $p_T > 3$ GeV/c, and all other events. The fraction of simulated events belonging to each of the six classes is adjusted to match the

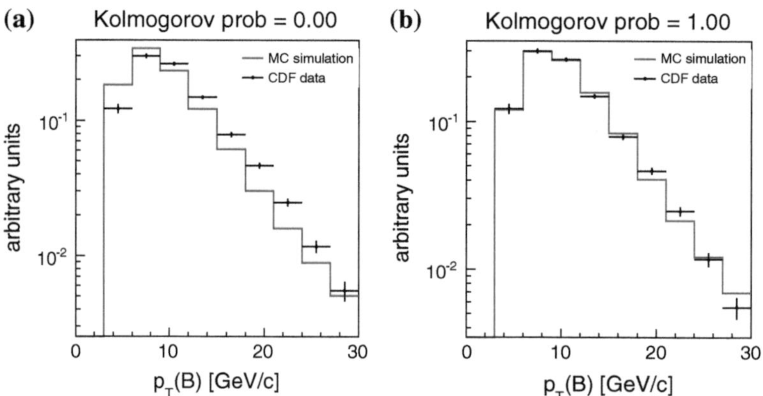

Fig. 5.1 Comparison between $p_T(B_s^0)$ distribution in data (*black points*) and simulation (*red histograms*) for non reweighted (**a**) and reweighted (**b**) simulation

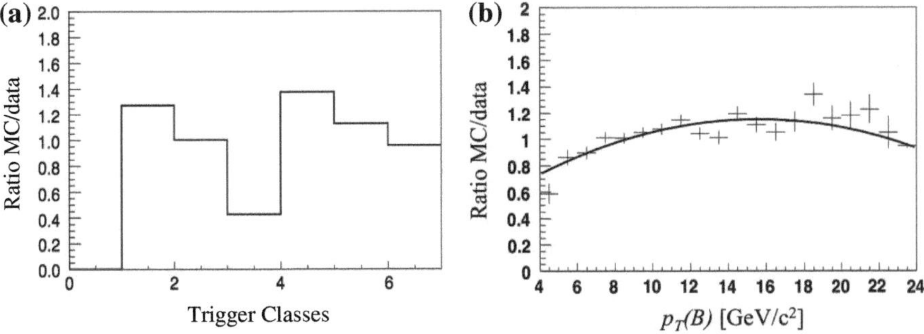

Fig. 5.2 Simulation weights. Discrete weights associated to the trigger classes (**a**). Bins 1–3 represent the weights associated to the three CMU–CMU trigger classes; bins 4–6 refer to CMU–CMX trigger classes. Continuous weight function, as resulting from a fit to the ratio of the $p_T(B)$ distributions of simulation and data (**b**)

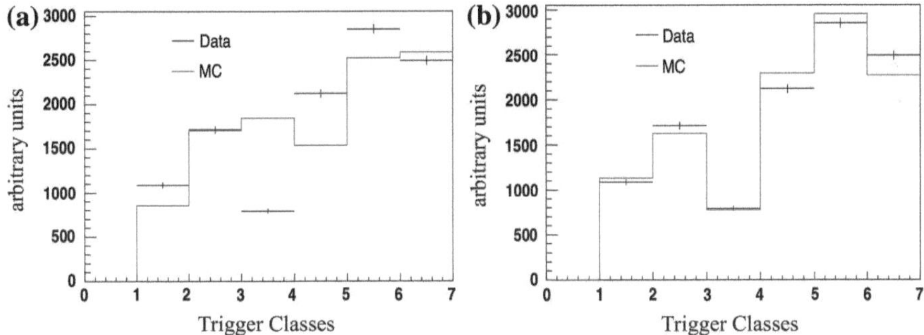

Fig. 5.3 Comparison between the data (*black points*) and phase-space simulated distributions (*red line*) in trigger categories for non-reweighted (**a**) and reweighted (**b**) simulation

fraction observed in data. The three classes are mutually exclusive with good approximation and are such that their union corresponds to the whole sample. A weight is obtained as the ratio between the number of real events belonging to one of the six classes over the corresponding number of simulated events. Figure 5.2a shows the resulting weights.

$p_T(B_s^0)$ **distribution:**   we compare the $p_T(B_s^0)$ distribution in the 4–24 GeV/c range between data and reweighted simulated events. The number of simulated events is normalized to the number of the signal events in the data. This ratio is fitted with a second order polynomial (Fig. 5.2b) function that is then used to reweight the simulated events.

Combined effect of trigger and $p_T(B_s^0)$ distribution: the final weight factor associated to each simulated event accounts simultaneously for the trigger-class weight factor and the $p_T(B_s^0)$ factor. Figures 5.3 and 5.1b show the final improved agreement between data and reweighted simulated events.

Having obtained simulated events that reproduce accurately data in all salient features relevant for angular distributions, we proceed to derive angular acceptances.

5.1.2 Angular Efficiency Fit Results

In order to obtain the coefficients a_{lm}^k in Eq. (5.1), the anglular distribution of reweighted simulated events are sampled using 20 bins in each dimension. The resulting angular sculpting is shown in Fig. 5.4. Assuming that the background is composed by mainly uncorrelated tracks, its angular distributions reconstructed in data should be flat. Any deviation from a constant distribution has to be attributed to detector sculpting. Figure 5.5 shows the angular distribution for $B_s^0 \to J/\psi\phi$ events in the mass sideband region, with shapes very similar to what obtained in simulations (Fig. 5.4). Furthermore, validation of detector sculpting has been obtained in data by performing angular analyses of $B^0 \to J/\psi K^\star$ decays [3, 4], obtaining results of polarization amplitudes compatible with results from B-factories [5, 6].

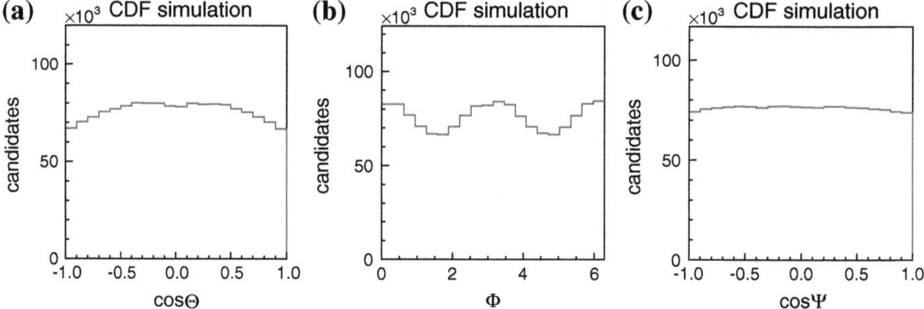

Fig. 5.4 Observed distribution of the transversity angles $\cos(\Theta)$ (**a**), Φ (**b**), and $\cos(\Psi)$ (**c**) for simulated events generated flat in the angular variable space

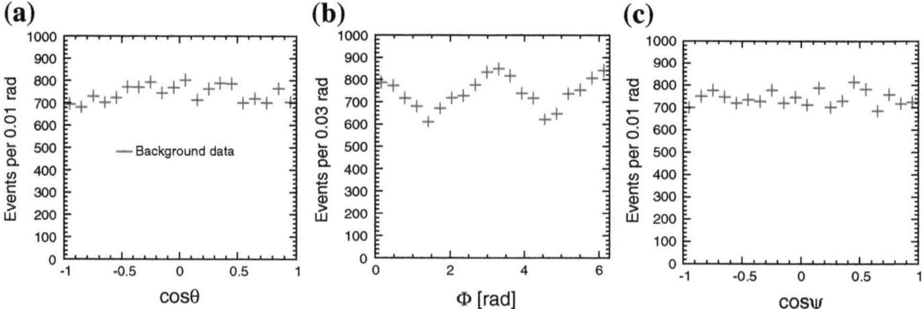

Fig. 5.5 Observed distribution of the transversity angles $\cos(\Theta)$ (**a**), Φ (**b**), and $\cos(\Psi)$ (**c**) for events in the mass sideband region $(5.291 < m_{B_s^0} < 5.315\,\mathrm{GeV/c}^2) \cup (5.417 < m_{B_s^0} < 5.442\,\mathrm{GeV/c}^2)$

Overall the angular sculpting has a moderate impact, with excursions of at most 15 %. However the full three-dimensional acceptance map is needed to correctly account for the expected angular correlations in $B_s^0 \to J/\psi\phi$ decays. The three-dimensional histogram is fitted with an expansion of real spherical harmonics for the (Θ, Φ) angles, where each term of the spherical harmonic series is further expanded as a function of Legendre polynomials [7] used to fit the Ψ angle. The fit to (Θ, Φ) is expanded in terms of real spherical harmonics according to the Laplace series

$$f(\Theta, \Phi) = \sum_{l=0}^{\infty} \sum_{m=0}^{l} [C_{lm} \cos(m\Phi) + S_{lm} \sin(m\Phi)] P_l^m (\cos\Theta), \qquad (5.2)$$

where the coefficients C_{lm} and S_{lm} are expanded as orthogonal Legendre polynomials functions such that

$$C_{lm} = \sum_{k=0}^{\infty} C_{lm}^k \sqrt{\frac{2k+1}{2}} P_k(\cos\Psi), \qquad (5.3)$$

$$S_{lm} = \sum_{k=0}^{\infty} S_{lm}^k \sqrt{\frac{2k+1}{2}} P_k(\cos\Psi), \qquad (5.4)$$

where C_{lm}^k and S_{lm}^k are the final coefficients that can be converted into the a_{lm}^k parameters as in Eq. (5.1). A two dimensional fit to the $(\cos\Theta{-}\Phi)$ distribution and the resulting residuals for all $\cos\Psi$ is shown in Fig. 5.6. Any potential mismodeling is properly included in the systematic uncertainties (Sect. 8.1).

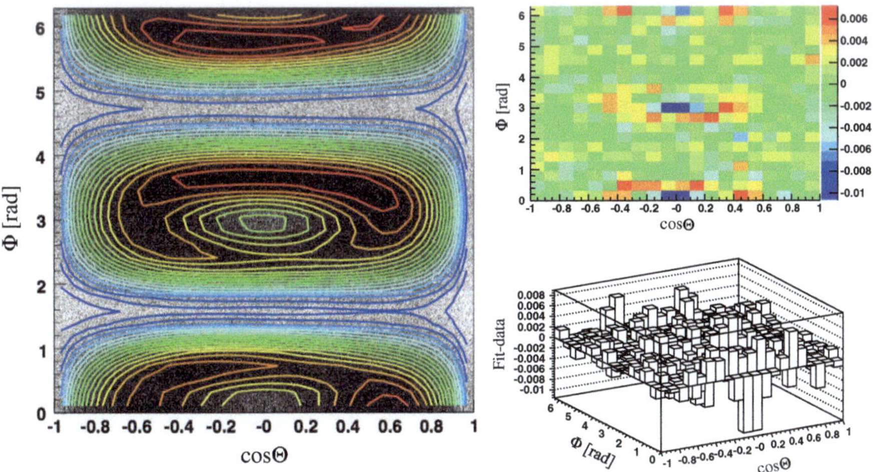

Fig. 5.6 Representation of the iso-likelihood lines in the two-dimensional angular space $(\cos\Theta, \Phi)$ as resulting from the fit to simulated events to parametrize the angular acceptance (**a**). Residuals of the fitted parametrization with respect to the original histogram (**b**)

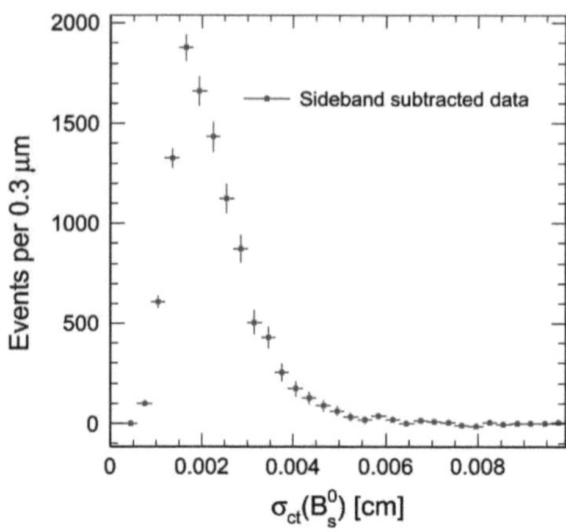

Fig. 5.7 Proper decay-time uncertainty distribution of B_s^0 candidates. Events are restricted to the $5.34 < m_{B_s^0} < 5.39\,\mathrm{GeV/c^2}$ region whose distribution is subtracted of sideband data distributions with $(5.291 < m_{B_s^0} < 5.315\,\mathrm{GeV/c^2}) \cup (5.417 < m_{B_s^0} < 5.442\,\mathrm{GeV/c^2})$

5.2 Decay-Time Resolution

Decay-time resolution is the single most important factor to resolve B_s^0 fast oscillations and enhance the sensitivity on the $\beta_s^{J/\psi\phi}$ mixing phase. In 2006, a large amount of dedicated studies were devoted to understand and optimize the decay-time resolution for the measurement of B_s^0 oscillations [8]. These resulted in an improved alignment of the tracking detector and the understanding of the per-event decay-time resolution up to a scalar correction factor. Figure 5.7 shows the decay-time uncertainty distribution as observed in the full CDF data sample for B_s^0 signal events populating the region $5.34 < m_{B_s^0} < 5.39\,\mathrm{GeV/c^2}$, from which the background distribution of events restricted to $(5.291 < m_{B_s^0} < 5.315\,\mathrm{GeV/c^2}) \cup (5.417 < m_{B_s^0} < 5.442\,\mathrm{GeV/c^2})$ is subtracted. The distribution peaks at $\sigma_{ct} \approx 65$ fs and is averaged at $\langle \sigma_{ct} \rangle \approx 78$ fs. By scaling this value for the known factor of 1.15, we obtain the final decay time resolution of 90 fs.

5.3 Flavor Identification

Information on the B_s^0 meson flavor at production is a key aspect of the analysis since the time evolution for particles and antiparticles is different: identifying the production flavor eliminates two of the four symmetries of the decay rate in Eq. (2.39) and enhances the sensitivity to the CP-violating mixing phase, by allowing access to additional terms proportional to the oscillation frequency as shown in Sect. 2.4.2.

The determination of the production flavor of B mesons is referred to as flavor tagging, and is achieved through algorithms called flavor taggers.

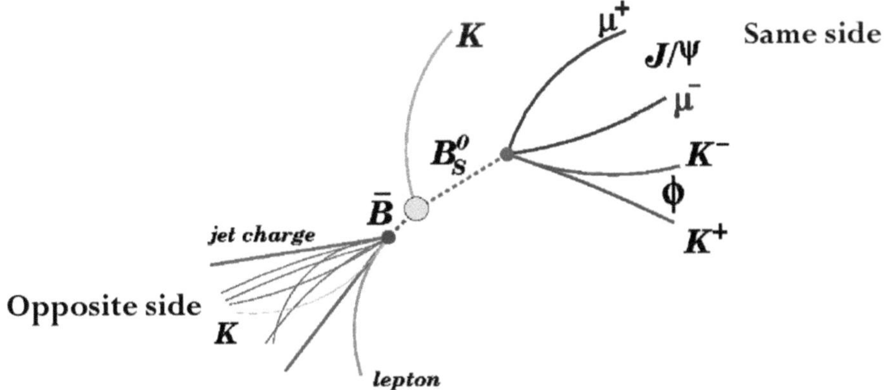

Fig. 5.8 Sketch of production and decay of a $b\bar{b}$ pair in $p\bar{p}$ collisions exemplifying the principles used in same-side and opposite-side flavor tagging

The properties of the $p\bar{p} \rightarrow b\bar{b}$ production process, and b quark hadronization and fragmentation are used in two general classes of flavor tagging algorithms employed at CDF: *opposite side* (OST) and *same side* (SST) taggers (Fig. 5.8).

Opposite side taggers use the fact that most b quarks produced in $p\bar{p}$ collisions are produced in $b\bar{b}$ pairs. The b or $\bar{b}$ quark produced along with the bottom-strange signal candidate hadronizes into another B hadron. The flavor of such *opposite side* B hadron can be tagged by identifying its fragmentation or decay products. If the opposite side hadron is tagged as a containing a b quark, the signal candidate is tagged as being produced as a B_s^0 meson. If the opposite-side hadron is tagged as containing a $\bar{b}$ quark, the signal candidate is tagged as being produced as a $\bar{B}_s^0$ meson. The anticorrelation between the opposite-side and signal flavor at production can be degraded if the opposite-side hadron is a neutral meson that undergoes oscillations before decaying. Same side taggers use charge-flavor correlations between the signal bottom-strange candidate and the fragmentation particles produced in its hadronization to determine its flavor.

Because of measurement uncertainties, acceptances, and approximations in the tagging algorithms, event information sufficient for a tag decision is not always available. As a result, a tagger is characterized by an efficiency ε, the fraction of times that it produces a tag decision,

$$\varepsilon = \frac{N_{tagged}}{N_{untagged} + N_{tagged}}, \tag{5.5}$$

where N_{tagged} is the number of signal B candidates for which a tag decision is available and $N_{untagged}$ is the number of B candidates for which no such decision is available. In addition, a fraction of the tagging decisions may be incorrect. The probability of a wrong-tag decision is denoted by ω, a *mistag probability* used to define a *dilution*, which is the asymmetry between correctly-(N_R) and wrongly-(N_W) tagged event yields

$$D \equiv 1 - 2\omega \quad \text{or equivalently} \quad D = \frac{N_R - N_W}{N_R + N_W}.$$ (5.6)

An ideal tagger that always makes the correct decisions has unit dilution. A fully random tagger has zero dilution. Hence, in the case of a pure signal sample comprising a number N of B mesons and $\bar{N}$ of $\bar{B}$ mesons, the particle-antiparticle asymmetry before tagging, $\mathcal{A} = (N - \bar{N})/(N + \bar{N})$, is modified by the mistag probability through the dilution yielding the observed asymmetry

$$\begin{aligned}
\mathcal{A}_{meas} &= \frac{n - \bar{n}}{\varepsilon(N + \bar{N})} \\
&= \frac{\varepsilon(1 - \omega)N + \varepsilon\omega\bar{N} - \varepsilon(1 - \omega)\bar{N} - \varepsilon\omega N}{\varepsilon(N + \bar{N})} \\
&= (1 - 2\omega)\mathcal{A} = D\mathcal{A}.
\end{aligned}$$ (5.7)

In CDF, the outcome of the tagging algorithm is parametrized for each candidate as a tag decision, ξ, an integer variable that assumes values ± 1 or 0, and a predicted dilution D. A value of $\xi = -1$ ($\xi = 1$) implies that the signal bottom-strange meson is tagged as being a B_s^0 ($\bar{B}_s^0$) at production. When the tagger is unable to reach a decision, the value is $\xi = 0$. The probability of a wrong-tag decision depends on several effects. In order to assign a larger (smaller) weight to events associated to a higher (lower) confidence in the tag decision, the dilution for each tagger is parametrized as a function of several relevant event variables, and a predicted dilution is assigned to each event as a result. Inclusion of the predicted dilution in the event observables enhances the overall tagger performance, and allows more accurate use of tagger calibrations from one event sample to another.

The figure of merit εD^2, where D is the dilution averaged over the studied sample, quantifies the overall performance of flavor tagging algorithms, providing a quantitative estimate of the effective reduction in signal-sample size when a correctly tagged signal sample is needed.

To use tagging information in the time-evolution fit we characterize the tagging algorithms performance. In the following we detail the calibration and use of OST and SST algorithms in this analysis.

5.3.1 Opposite Side Tagging

Different OST algorithms have been developed in CDF. A detailed description of them is beyond the scope of this thesis. Only a brief overview of the working principles is given. Further information is available in Refs. [9–11].

5.3.1.1 Lepton Taggers

Lepton taggers infer the production flavor of the signal meson from the semileptonic decay of the opposite-side B-hadron through the transitions $b \to cl^-\bar{\nu}_l X$ and $\bar{b} \to \bar{c}l^+\nu_l X$. Identification of a positively-charged lepton (muon or electron) indicates a $\bar{b}$ content for the tagging hadron, while a negatively-charged lepton indicates a b quark. Two algorithms are developed, the soft muon tagger, and the soft electron tagger [9]. The typical values of semileptonic branchings fraction $\mathcal{B}(B \to lX) \approx 20\,\%$ result in limited efficiency. The efficiency is further reduced by leptons that may not be detected in the fiducial volume of the acceptance. The accuracy of the tag decision is affected by several factors. Semileptonic decays of $B^0_{(s)}$ mesons may happen after flavor oscillations occurred, resulting in a wrong tag. In the case of semileptonic B^0_s decays, the fast oscillations almost nearly randomize the lepton tag. If the identified lepton comes from a $b \to c \to lX$ transition, the inferred flavor will be opposite to the true flavor state. The good purity of lepton identification at CDF limits the impact of lepton misidentification. Lepton identification is achieved using multivariate likelihood discriminants that combine information from several subdetectors into a simple scalar variable that offer strong discrimination between real leptons and other particles misidentified as leptons. Prior to use in lepton taggers candidates are required to have $p_T > 4$ GeV/c, hits in CMP and CMU detectors for muons, and energy deposition $E_T > 4$ GeV in the electromagnetic calorimeter for electrons. Lepton selection criteria include also track quality requirements with at least $4\ r - \phi$ hits in silicon detector and impact parameter $0.12 < d_0 < 1$ mm.

The performance of the lepton taggers shows dependencies on two variables: the value of lepton likelihood and the transverse momentum of the lepton with respect to the axis of the hadronic jet in which the lepton is reconstructed, denoted by p_T^{rel}. The dilution improves with higher lepton likelihood, as expected since a higher purity of the lepton sample yields better b flavor determination. The dilution also improves with higher p_T^{rel} because leptons from the decay of a heavy hadron tend to have a larger momentum spread in the plane transverse to the b jet than those in jets coming from lighter quarks. The dependence on p_T^{rel} is determined empirically in samples of events enriched in leptons and displaced tracks [12]. Hence, the dilution of the soft lepton tagger for each candidate is predicted based on the observed values of lepton likelihood and p_T^{rel}.

5.3.1.2 Jet Charge Taggers

Jet-charge tagging uses the difference in electric charge between b and $\bar{b}$ quarks to identify the bottom-quark content of the signal meson. Jet charge taggers estimate the net charge of the hadronic jet recoiling against the signal B^0_s [10]. The net charge is estimated by the momentum-weighted sum of the charges particles within the considered jet. Jets are identified and reconstructed by the *cone clustering algorithm* [13] on the hemisphere opposite to the direction of the signal meson. This simple geometrical algorithm starts with an high-p_T track seed, to which low-p_T

tracks in an isolation cone of $\Delta R < 0.7$ with respect to the seed track are added. Selection requirements on $|d_0|$, z_0 and p_T are imposed on the tracks to be included in jet reconstruction. Finally, artificial neural networks (NN) are used to identify the jet that most likely originates from a b quark among the set of reconstructed jets. The NNs are trained and optimized on large samples of simulated events. The identification of the highest probability bottom jet involves two steps. A first NN, using as input several kinematic variables, is used to assign to each track a probability P_{trk} of having originated from the decay of a B hadron. A second NN assigns to each jet a probability P_{jet} of deriving from a b quark, using kinematic variables and the probability P_{trk} for each track. Once the highest probability bottom jet is found, the jet charge is computed as

$$Q_{jets} \equiv \frac{\sum_i Q^i p_T^i (1 + P_{trk}^i)}{\sum_i p_T^i (1 + P_{trk}^i)}, \tag{5.8}$$

where the sum runs over all the tracks of the chosen jet, and Q^i, p_T^i, and P_{trk}^i are respectively the charge, transverse momentum, and probability of having originated from a B decay for the ith track. In order to exploit the greater statistical power available from jets with different features, this procedure is optimized for three exclusive classes of jets. Class 1 jets contain an identified secondary vertex with a decay-length significance $L_{xy}/\sigma_{L_{xy}} > 3$. Class 2 jets do not fulfill the criteria for Class 1 and have at least one track with $P_{trk} > 0.5$. Class 3 jets are the remaining jets that pass the selection criteria. Such criteria provide a very inclusive sample of jets achieving 50 % tagging efficiency, at the price of reduced dilution, since the heavy flavor content of the considered jets is limited. The dilution for jet-charge tagging is linearly parametrized as a function of the quantity $|Q_{jet}|p_{jet}$, where Q_{jet} and p_{jet} are the net charge and the momentum vector of the jet. The parametrization is done separately for the three jet classes.

5.3.1.3 Opposite Side Kaon Tagger

This algorithm exploits the correlation between the flavor of the opposite-side bottom hadron and the electric charge of the kaon originated from its $b \rightarrow c \rightarrow s$ decay. A b quark is identified by the presence of a K^- meson and a $\bar{b}$ quark by a K^+ meson. The algorithm searches for a track originating from a decay vertex suitably displaced from the primary vertex and tries to identify it as a kaon by using combined particle identification information from dE/dx and TOF to cover the widest momentum range possible. The opposite-side kaon tagger was developed using a sample enriched in semileptonic B decays selected by a trigger on events containing a lepton accompanied by a displaced track. The trigger used the silicon vertex tracker informations [14] to collect events coming from semileptonic decays of B mesons, such as $B \rightarrow D^{(*)} l \nu X$, by requiring a lepton candidate in the EM calorimeters or in the muon chambers accompanied by one displaced track. Basic quality selection requirements and requirements on $|d_0|$, z_0 and p_T are imposed to suppress the contribution of charged particles originating in material interactions or from decays of long-lived hyperons. The track is required to be associated to valid TOF and dE/dx information, combined in a likelihood to maximize the particle identification performance. The p_T spectrum of the tagging candidates tracks peaks around 1 GeV/c, a regime in which the larger contribution to K/π discrimination comes from the TOF detector. However, requiring TOF information reduces considerably the tagging efficiency, as TOF inefficiency can reach 50 % per track, depending on instantaneous luminosity conditions.

5.3.1.4 Combination of Opposite Side Taggers

Tagging decisions may not be independent, since various algorithms can use the same tracks ar different tracks that are kinematically correlated. Any correlation needs to be accounted for when using the three taggers together in a fit. A NN combination of taggers yields a single opposite-side tagging decision [15] that properly includes correlations, and achieves a factor 15 % improvement over simple linear combinations. The NN is trained on data selected by a trigger on a lepton and a displaced track. Relevant inputs to the NN are the tag decisions and predicted dilutions of the individual taggers and several kinematic quantities of the event. The muon tagger provides the single most relevant contribution to the OST decision. Since all OST processes are independent from the signal candidate hadronization, the same opposite side tagging algorithm can be applied to samples of different B meson species. The OST algorithms are developed, calibrated, and optimized with copious samples of charged B meson decays in which the true flavor is known, and then applied to B_s^0 decays.

5.3.2 OST Calibration

Since the signal decays used in this analysis are reconstructed in samples collected by different triggers from those used to collect the OST-calibration samples, we verify that the predicted dilution is not sample dependent, and that the dilution measured on dimuon data, $D^{\mu\mu}$, matches the predicted dilution D^{lt} from calibration on "lepton + displaced track" data. We assume that any discrepancy can be parametrized with sufficient accuracy by a linear dependence and allow for a multiplicative scale factor such that $D^{\mu\mu} = S_D D^{lt}$. If the parametrization of the dilution is a good description of this signal sample, the scale factor S_D as determined in data should be consistent with unity.

The determination of the scale factors appropriate for the full Run II data sample is achieved by applying the OST algorithms to a large sample of fully reconstructed

$B^+ \rightarrow J/\psi K^+$ decays collected by the same trigger as the signal sample. The tagging decision and associated predicted dilution of the algorithm is then compared to the actual bottom-quark content of the meson, which is known from the charge of the kaon. Because charged B mesons do not oscillate, the kaon charge identifies the flavor at production. The $B^+ \rightarrow J/\psi K^+$ decay also benefits from higher yield than the B^0_s samples, with about 81,000 signal events, allowing the study of any mismodelling with sufficient accuracy.

5.3.2.1 $B^+ \rightarrow J/\psi K^+$ Calibration Sample

The $B^+ \rightarrow J/\psi K^+$ decays are reconstructed from a $p\bar{p}$ sample corresponding to an integrated luminosity of $\mathcal{L} \approx 10\,\mathrm{fb}^{-1}$ collected by the same dimuon trigger used to collect the signal sample. The offline selection is as similar as possible to the $B^0_s \rightarrow J/\psi \phi$ signal selection. We use preselection criteria including the following requirements

- $5.16 < M(J/\psi K^+) < 5.40\,\mathrm{GeV/c^2}$;
- ≥ 3 axial hits per track in the silicon detector;
- ≥ 3 axial and ≥ 3 stereo hits per track in the COT detector;
- successful XFT-muons match;
- decay-time uncertainty $0.0 < \sigma_{ct} < 0.1\,\mu\mathrm{m}$.

We then apply a neural network discriminant similar to the NN used for selecting $B^0_s \rightarrow J/\psi \phi$ decays (Sect. 4.3.2) but adapted to the different topology of the calibration sample. The selection requirement on the neural network output (NN > 0.8) is chosen by optimizing $S/\sqrt{S+B}$ [16], where S and B are the number of signal and background events in the mass signal region as reconstructed in simulated events, respectively. The signal region is defined to be within about $\pm 2.5\sigma$ of the known B^+ mass, $5.24 < m_{B+} < 5.32\,\mathrm{GeV/c^2}$, while the sideband regions are chosen to be in the 3σ–6σ range away from the peak, $(5.17 < m_{B+} < 5.21) \cup (5.35 < m_{B+} < 5.39)\,\mathrm{GeV/c^2}$. Alterate selection criteria are studied in detail in this work, aiming at not only improving signal-to-background separation but also at optimizing the tagging performances in terms of unbiasedness and accuracy. We attempted several variations of the selection criteria commonly used to select the $B^+ \rightarrow J/\psi K^+$ decays, involving lifetime-related, kinematic, and PID-related quantities. As a result of this systematic exploration, we found that a lower threshold on the decay length at $60\,\mu\mathrm{m}$, which rejects a large fraction of combinatorial background, while preserving about $85\,\%$ of signal, is very effective in ensuring a more robust tagger performance. The resulting $B^+ \rightarrow J/\psi K^+$ sample (Fig. 5.9) has an higher purity than previously achieved, which allows an increased consistency between the calibration on the B^+ and B^- sample. A simple Gaussian fit over a linear background determines approximately 40,000 B^- decays and 41,000 B^+ decays. We neglect small contribution from misreconstructed $J/\psi \pi^+$ decays since these have no impact in the calibration.

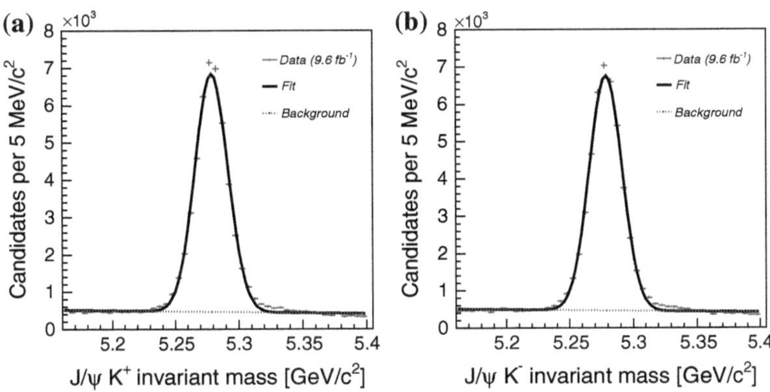

Fig. 5.9 $J/\psi K^+$ mass distribution (**a**) and $J/\psi K^-$ mass distribution (**b**) for the full Run II data sample

5.3.2.2 Scale Factors Determination

To characterize the OST performance, for each decay we compare the true flavor (as indicated by the kaon charge) with the flavor identified by the OST algorithm. This is done by comparing the known dilution (since we know the B^+ flavor exactly) with the dilution predicted by the algorithm. We divide the sample in independent subsamples according to their predicted dilution. For each range of predicted dilution we determine the number of right (wrong) tags to extract the actual dilution. Then we display the actual dilution as a function of the predicted one (Fig. 5.10) to determine the scale factor through a linear fit. All dilution distributions are background-subtracted using mass sidebands. We investigate independent B^+ and B^- scale factors, in order to allow for any asymmetry in the tagging algorithms.

We find $S_D^+ = 1.09 \pm 0.05$ and $S_D^- = 1.08 \pm 0.05$, respectively, with a total average dilution of $\overline{D} = 6.88 \pm 0.03\,\%$. We also determine separately the scale factors for the first $5.2\,\mathrm{fb}^{-1}$ of data (Fig. 5.11) and the latest $4.4\,\mathrm{fb}^{-1}$ of data (Fig. 5.12) and compare them to previous results of this analysis to identify any change in performance. The results are summarized in Table 5.1. The small differences in scale factors with respect to the previous analysis are compatible with the changes in sample composition due to our improved selection. Table 5.2 reports a comparison of the scale factors determination with and without the selection improvements using the whole data set.

The new selection directly reflects into an $\mathcal{O}(30\,\%)$ improvement in precision of the scale factor determination. Since the calibrated values of the scale factor for B^+ and B^- mesons are very consistent, we average them in the fit, using a single scale factor, and its spread as a systematic uncertainty. This reduces the number of nuisance parameters with respect to previous iterations of the analysis where flavor-specific scale factors were used. The improved accuracy in the scale-factor determination,

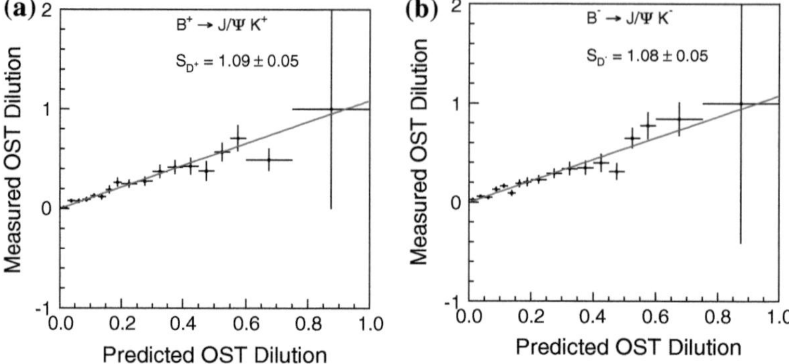

Fig. 5.10 Measured dilution as a function of predicted dilution for B^+ (**a**) and B^- decays (**b**) for all data

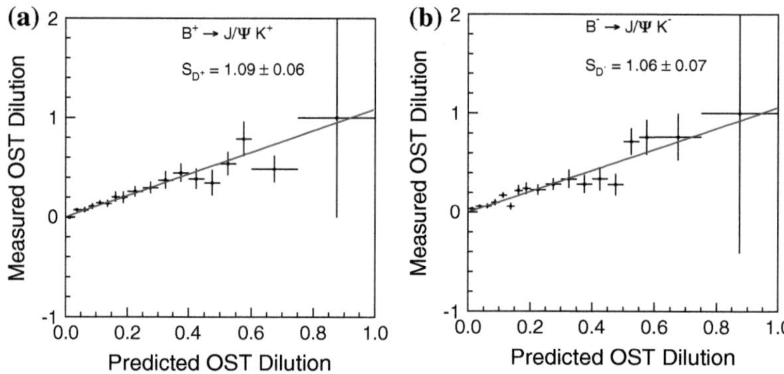

Fig. 5.11 Measured dilution as a function of predicted dilution for B^+ (**a**) and B^- decays (**b**) for data corresponding to 5.2 fb^{-1}

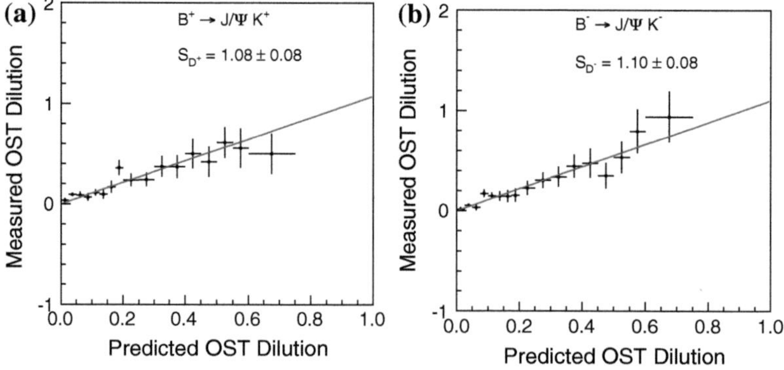

Fig. 5.12 Measured dilution as a function of predicted dilution for B^+ (**a**) and B^- decays (**b**) for the last 4.4 fb^{-1} of data

Table 5.1 OST performance for B^+ and B^- decays in different parts of the data, compared with values from the previous analysis [17]

Scale factor	Reference [17] (0–5.2 fb^{-1})	0–5.2 fb^{-1}	5.2–10 fb^{-1}	0–10 fb^{-1}
$S_D{}^+$	0.93 ± 0.09	1.09 ± 0.06	1.08 ± 0.08	1.09 ± 0.05
$S_D{}^-$	1.12 ± 0.10	1.06 ± 0.07	1.10 ± 0.08	1.08 ± 0.05
ε	94.3 ± 0.3	93.8 ± 0.1	91.2 ± 0.2	92.8 ± 0.1
$\overline{D}$ (%)	6.9 ± 0.1	6.84 ± 0.04	6.93 ± 0.05	6.88 ± 0.03
$\sqrt{\overline{D^2}}$ (%)	11.5 ± 0.02	11.26 ± 0.08	11.36 ± 0.10	11.30 ± 0.06
εD^2 (%)	1.2	1.19 ± 0.02	1.18 ± 0.02	1.18 ± 0.01
$\varepsilon S D^2$ (%)		1.30 ± 0.06	1.29 ± 0.07	1.39 ± 0.05

Table 5.2 OST scale factor results determined in the whole data set with, and without the $c\tau > 60\,\mu$m requirement

	No $c\tau$ requirement	$c\tau > 60\mu$ m
$S_D{}^+$	0.95 ± 0.05	1.09 ± 0.05
$S_D{}^-$	1.13 ± 0.05	1.08 ± 0.05

and consistency between B^+ and B^- samples is one of the relevant improvements of this analyses with respect to previous iterations.

As an additional consistency-check, we determine scale factors and efficiencies as functions of time, to ensure stability and consistency throughout the whole data set. Figure 5.13 shows that scale factors are stable in time. A tendency towards decreased tagging efficiency, correlated with higher instantaneous luminosity is observed in later data (Fig. 5.14). The decrease in average efficiency is less than 3 % and impacts only marginally the resolution on the extracted physical parameters, while is not expected to bias their values.

5.3.2.3 Opposite Side Tagging Results

The OST scale factor measured in the whole CDF dataset is $S_D = 1.09 \pm 0.05$, with a tagging efficiency of $(92.8 \pm 0.1)\%$ and a mean predicted dilution of $(6.88 \pm 0.03)\%$. The effective opposite-side tagging power is $\varepsilon(S_D D)^2 = (1.39 \pm 0.05)\%$.

5.3.3 Same Side Kaon Tagger

The working principle of the same-side-tagging method significantly differs from that of opposite-side tagging, where the hadronization and fragmentation of the opposite-side b quark is independent of the species of the B meson of interest. In same-side

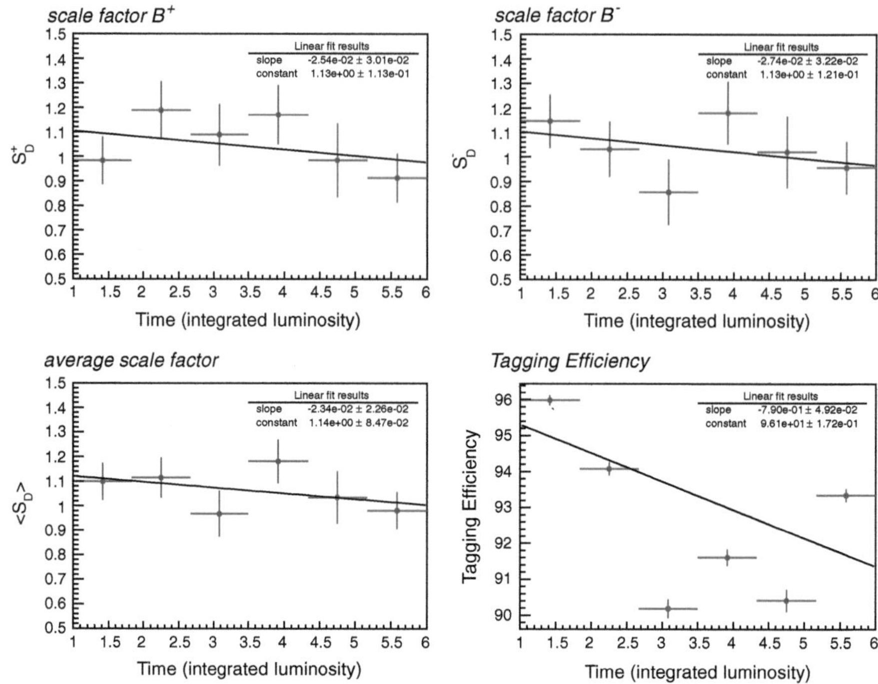

Fig. 5.13 OST performances aa functions of time. Dilution scale factors for B^+ decays (*top, left*), B^- decays (*top, right*), their average (*bottom, left*) and efficiency (*bottom, right*)

tagging, the flavor tag is determined using particles that are produced in association with the signal bottom-strange meson. Since the species of particles produced in association with the signal meson vary depending on B meson species, the performance is necessarily dependent on the signal meson species and its kinematic distributions. The charged particles associated with the signal B meson arise from light quarks that accompany the hadronization process or in the decays of excited states $B^{**} \to Bh$ in the strong-interaction decays of orbitally-excited B^{**} states produced in the $p\bar{p}collision$; h could be a π or a K depending on the B-hadron flavor. Thus, the flavor of the light quark in the signal meson is correlated with the species of associated particles produced in the first hadronization branches, which are in principle different for a B^+, a B^0, or a B_s^0 meson. In the case of the B_s^0 meson, the leading associated track is a kaon. A K^+ meson tags a B_s^0 meson, while a K^- meson tags a $\bar{B}_s^0$ meson. However, the B_s^0 meson could be accompanied by a neutral strange particle, as K^{*0} or Λ, which would prevent to tag the B_s^0 flavor and therefore decrease the tagging efficiency. The quality selection of tracks used for same-side tagging is similar to the selection for opposite side tracks. Tracks must be within a cone around the B_s^0 momentum direction to be considered tag candidates, and have $p_T > 0.4$ GeV/c. The same-side-tagging power performance benefits from particle identification (dE/dx and TOF) to identify the associated track as a kaon.

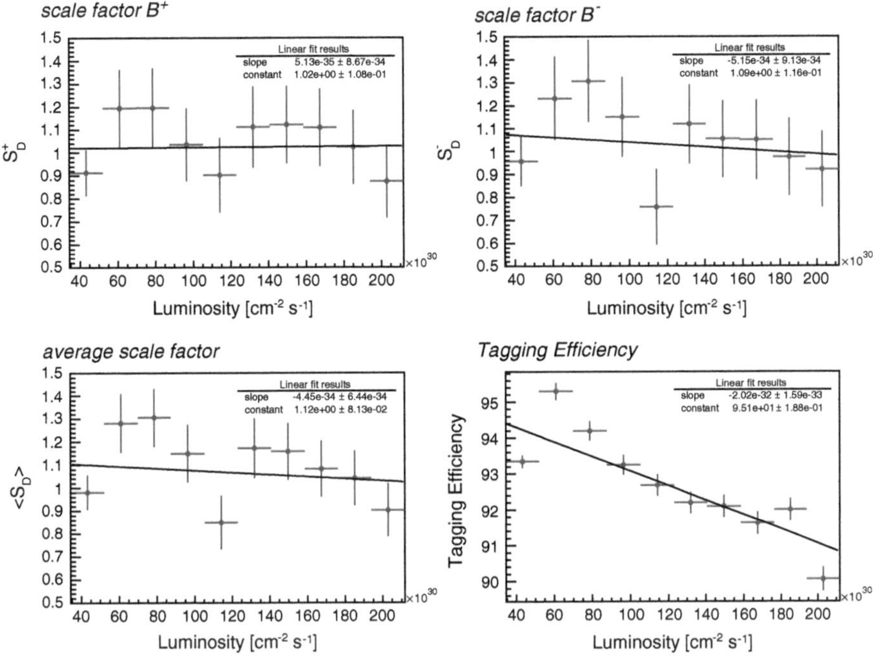

Fig. 5.14 OST performances as functions of istantaneous luminosity. Dilution scale factors for B^+ decays (*top, left*), B^- decays (*top, right*), their average (*bottom, left*) and efficiency (*bottom, right*)

This decreases the mistag rate, which may occur if a pion is mistakenly identified as an associated kaon. Particle identification, as well as information about track momentum, is used to decide which track to choose as the tagging track, if multiple possible tagging tracks are in the event.

5.3.3.1 SSKT Calibration

The same-side kaon-tagger (SSKT) algorithm was developed on samples of copious simulated B_s^0 decays and calibrated on B_s^0 data for the CDF B_s^0 oscillation frequency measurement [8, 18]. The version used in this analysis was calibrated on 5.2 fb^{-1} of data by repeating the measurement of the B_s^0 mixing and extracting the dilution [17]. The calibration uses four decays collected using a trigger on displaced tracks:

- $B_s^0 \rightarrow D_s^- \pi^+, \; D_s^- \rightarrow \phi(1,020)\pi^-, \; \phi(1,020) \rightarrow K^+K^-$;
- $B_s^0 \rightarrow D_s^- \pi^+, \; D_s^- \rightarrow K^*(892)^0 K^-, \; K^*(892)^0 \rightarrow K^+\pi^-$;
- $B_s^0 \rightarrow D_s^- \pi^+, \; D_s^- \rightarrow \pi^+\pi^+\pi^-$;
- $B_s^0 \rightarrow D_s^- \pi^+\pi^+\pi^-, \; D_s^- \rightarrow \phi(1,020)\pi^-, \; \phi(1,020) \rightarrow K^+K^-$.

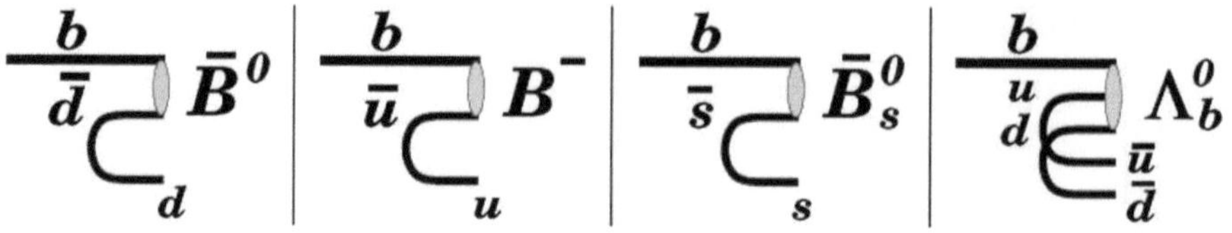

Fig. 5.15 Sketch of the first branch of B hadron fragmentation

where about half of the signal events are contributed by the first channel. At Level 1 the trigger reconstructs tracks in the drift chamber. It requires two oppositely-charged particles with thresholds in individual and combined transverse momenta. At Level 2, a dedicated custom-hardware processor associates the $r - \phi$ position of hits in the silicon detector with the reconstructed tracks [14]. This provides a precise measurement of the track impact parameter, d_0, which is used to efficiently discriminate hadronic decays of heavy flavor particles from light-quark backgrounds. A complete event reconstruction is performed at Level 3. Offline reconstruction of B_s^0 meson candidates begins by selecting D_s^- meson candidates. The $\phi(1{,}020)$ and $K^\star(892)^0$ intermediate resonances in the D_s^- decays are exploited to suppress background, by requiring candidates to be consistent with their known masses [19]. The D_s^- meson candidates are then combined with one or three additional charged particles to form a B_s^0 candidate. We constrain the D_s^- momentum and the additional particle(s) to originate from a common vertex in three dimensions.

The probability for observing the B_s^0 in a flavor eigenstate as a function of time is

$$P(t)_{B_s^0(\bar{B}_s^0)} \propto |1 \pm \cos \Delta m_s t|. \tag{5.9}$$

By folding the effect of the SSKT dilution, the probability becomes

$$P(t)_{B_s^0(\bar{B}_s^0)} \propto |1 \pm D \cos \Delta m_s t|. \tag{5.10}$$

By introducing an amplitude, A, such that

$$P(t)_{B_s^0(\bar{B}_s^0)} \propto |1 \pm A D_p \cos \Delta m_s t|, \tag{5.11}$$

the mixing frequency Δm_s is measured through an *amplitude scan* [20]. The parameter Δm_s is fixed to different probe values, and a fit for the amplitude is performed at each point. If the same side tagging is properly calibrated, the amplitude is consistent with unity when Δm_s approaches the true value and with zero elsewhere. An over- (under-)estimation of the dilution implies a measured amplitude value smaller (greater) than one.

Figure 5.16 shows the results of the amplitude scan on $5.2\,\mathrm{fb}^{-1}$ of data [17]. The amplitude is consistent with unity at $\Delta m_s = 17.79 \pm 0.07\,\mathrm{ps}^{-1}$, a value consistent with the known values [19], showing that the SSKT is properly calibrated. The size

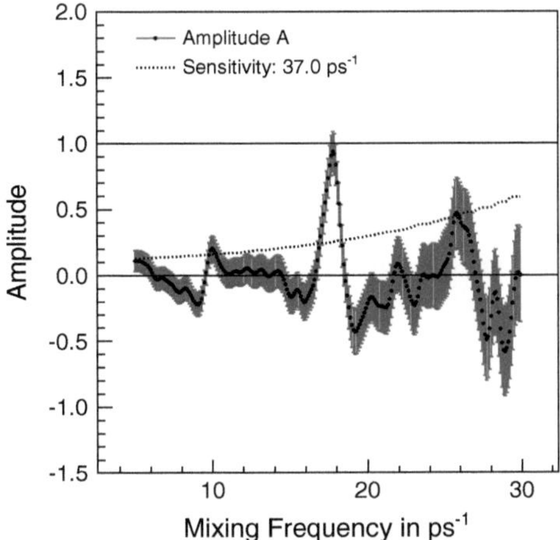

Fig. 5.16 Amplitude scan as function of Δm_s using $5.2\,\mathrm{fb}^{-1}$ of data to determine the SSKT dilution scale factor [17]

of the amplitude at maximum and the measured dilution scale factor for the SSKT yields $S_D = 0.94 \pm 0.2$. We measure a tagging efficiency of $(52.2 \pm 0.7)\%$, and an average predicted dilution on hadronic B_s^0 signal of $(25.9 \pm 5.4)\%$. The total SSKT tagging power is $(3.5 \pm 1.4)\%$.

5.3.3.2 Use of SSKT in the Analysis

We do not extend the calibration of the SSKT to the full data set. This is in mostly due to the marginal increase in calibration sample event yield with respect to the 5.2 fb^{-1} analysis. Only an approximately 20% increase in $B_s^0 \to D_s^- \pi^+$ signal sample size is available against an almost doubled integrated luminosity. The displaced-track trigger rate, which selects the hadronic decays used for calibrations, was severely suppressed in the latest part of data-taking period to prioritize high-p_T triggers aimed at collecting Higgs-boson decays. The lack of sufficient SSKT calibration data for the most recent data poses a significant challenge to finalizing the $\beta_s^{J/\psi\phi}$ measurement on the full CDF data set. In order to decide the best strategy, a number of detailed studies on simulated experiments are conducted. First, we evaluated the impact of the scale-factor resolution on the relevant fit parameters in the ideal case of SSKT scale factor known with infinite precision ($\sigma = 0$). Table 5.3 report the average fit results on about 1,000 simulated experiments, each corresponding to 5.2 fb^{-1} of data.

A better knowledge of the SSKT scale factor affects only marginally the estimated parameters uncertainties in our experimental conditions. We also estimate the impact of not using the SSKT in latest data, using simulated experiments. We find the impact of this choice to be limited. The average improvement in $\beta_s^{J/\psi\phi}$ resolution from including the SSKT in the full sample with a dilution known to the same level of accuracy as for the 5.2 fb^{-1} analysis does not exceed 12%, as shown in Fig. 5.17.

Table 5.3 Fit results on simulated experiments with different SSKT scale factor uncertainties

Fit variable	Pseudoexperiment results		Resolution improvement %
	(a)	(b)	
$\beta_s^{J/\psi\phi}$ [rad]	0.140 ± 0.128	0.138 ± 0.125	2.34
$\Delta\Gamma_s$ [ps^{-1}]	0.0686 ± 0.0289	0.0685 ± 0.0289	0
$c\tau$ [μm]	458.114 ± 6.141	458.08 ± 6.135	4.26

Column (a): SSKT scale factor Gaussian constrained to calibrated uncertainties, $S_D(\text{SSKT}) = 0.94 \pm 21.1\%$ [21]

Column (b): SSKT scale factor assumed to be known with infinite precision, $\sigma_{S_D(\text{SSKT})} = 0$. Last column report the percentage of improvement on the fit parameters resolution

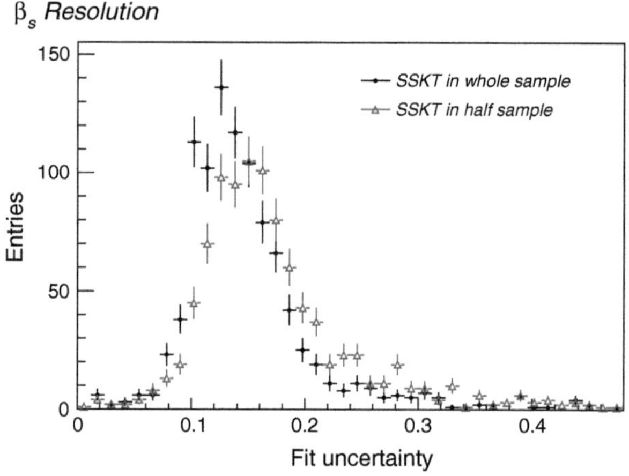

Fig. 5.17 Mixing phase resolution studies. *Black* (*blue*) *dots* refer to the uncertainty on the mixing phase estimated from the fit to simulated samples in which SSKT is used in the whole (first half of) data sample

Since recent data may suffer from a considerably larger uncertainty on SSKT dilution, the gain is likely to be even smaller.

Hence, we decided to use the calibration obtained with 5.2 fb^{-1} of data [21] and, accordingly, use the information from the SSKT tagger only in the first half of the sample. This conservative choice prevents potential systematic uncertainties arising from changes in the tagger performances as a function of time and does not significantly degrade the final sensitivity to the mixing phase.

References

1. M. Milnik, *Measurement of the lifetime difference and cp-violating phase in $B_s \rightarrow J/\psi\phi$ decays*, Ph.D. thesis, FERMILAB-THESIS-2007-38, 2007
2. F. Azfar et al., *An Update of the Measurement of the CP-Violating Phase β_s using Bs to $J/\psi\phi$ Decays*. CDF note 10053

3. CDF Collaboration, *Angular Analysis of $B_s^0 \to J/\psi\phi$ and $B^0 \to J/\psi K^{*0}$ Decays and Measurement of $\Delta\Gamma_s$ and ϕ_s*. CDF Note 8950 (2007)

4. D. Tonelli (CDF Collaboration), *Search for New Physics in the B_s^0 Mixing Phase* (2008), arXiv:0810.3229 (hep-ex)

5. B. Aubert et al. (BaBar), Measurement of CP violating asymmetries in B^0 decays to CP eigenstates. Phys. Rev. Lett. **86**, 2515 (2001), hep-ex/0102030

6. R. Itoh et al. (Belle Collaboration), Studies of CP violation in $B^0 \to J/\psi K^{*0}$ decays. Phys. Rev. Lett. **95**, 091601 (2005), arXiv:hep-ex/0504030 (hep-ex)

7. J. Boudreau, J.P. Fernandez, K. Gibson, G. Giurgiu, G. Gomez-Ceballos, L. Labarga, C. Liu, P. Maksimovic, M. Paulini, *Measurement of the CP Violation Paramter beta' in Bs to Jpsi Phi*. CDF note 8960

8. A. Abulencia et al. (CDF Collaboration), Observation of $B_s^0 - \bar{B}_s^0$ oscillations. Phys. Rev. Lett. **97**, 242003 (2006), arXiv:hep-ex/0609040 (hep-ex)

9. G. Giurgiu et al., *Updated Likelihood Muon tagger*. CDF Note 7644 (2005)

10. G. Bauer et al., *Improved Jet Charge Tagger for Summer Conferences 2004*. CDF Note 7131 (2004)

11. G. Salamanna et al., *Opposite Side Kaon Tagging*. CDF Note 8179 (2006)

12. G. Giurgiu, V. Tiwari, M. Paulini, J. Russ, *Muon B Flavor Tagging—A Likelihood Approach*. CDF note 7043 (2004)

13. M. Bosman et al., Jet finder library: version **1** (1998)

14. B. Ashmanskas et al. (CDF-II Collaboration), The CDF silicon vertex trigger. Nucl. Instrum. Meth. **A518**, 532 (2004), arXiv:physics/0306169 (physics)

15. CDF Collaboration, *Combined Opposite Side Flavor Taggers*. CDF Note 8460 (2006)

16. T. Aaltonen et al. (CDF Collaboration), First flavor-tagged determination of bounds on mixing-induced *CP* violation in $B_s^0 \to J/\psi\,\phi$ decays. Phys. Rev. Lett. **100**, 161802 (2008), arXiv:0712.2397 (hep-ex)

17. T. Aaltonen et al. (CDF Collaboration), Measurement of the *CP*-violating phase $\beta_s^{J/\psi\phi}$ in $B_s^0 \to J/\psi\,\phi$ decays with the CDF II detector. Phys. Rev. **D85**, 072002 (2012), arXiv:1112.1726 (hep-ex)

18. Bs Mixing Group, *Optimization of the Same Side Kaon Tagging Algorithm Combining PID and Kinematic Variables*. CDF note 8344

19. J. Beringer et al. (Particle Data Group), Review of particle physics (RPP). Phys. Rev. **D86**, 010001 (2012)

20. H. Moser, A. Roussarie, Mathematical methods for B^0 anti-B^0 oscillation analyses. Nucl. Instrum. Meth. **A384**, 491 (1997)

21. M. Feindt, T. Kuhr, M. Kreps, J. Morlock, A. Schmidt, *Public Note for the Calibration of the Same Side Kaon Tagger Using Bs Mixing*. CDF Note 10108 (2006)

Chapter 6
The Fit to the Time Evolution

This chapter describes the fit to the time-evolution of the flavor-tagged, CP-even and CP-odd fraction of $B_s^0 \to J/\psi\phi$ decays used in the measurement.

6.1 Generalities

We use the maximum likelihood method (ML) to determine the set of unknown parameters ϑ. Each of the N decay candidates is described by a vector of event observables x_i. We denote by $p(x|\vartheta)$ the probability density function (PDF) that describes the expected distribution of events in the space of event observables x, given the vector of parameters ϑ. We construct the likelihood function as

$$\mathcal{L}(\vartheta) = \prod_{i=1}^{N} p(x_i|\vartheta), \tag{6.1}$$

which is maximized with respect to ϑ. In practice we perform a numerical minimization of $-\ln \mathcal{L}(\vartheta)$. This is convenient to avoid finite precision problems that may arise when many small numbers are multiplied together. The numerical minimization is performed by using the MINUIT package [1] and the ROOT analysis framework [2].

6.2 Likelihood Function

The PDF for an event is the linear combination of two components, the signal and the background PDF. Each is normalized to unit and it is decomposed in products of PDFs for each event observable when event variables are independent, or includes multi-observable joint PDFs otherwise. We use the following event variables: the

© Springer International Publishing Switzerland 2015
S. Leo, *CP Violation in $B_s^0 \to J/\psi\phi$ Decays*, Springer Theses,
DOI 10.1007/978-3-319-07929-5_6

mass of B_s^0 candidates and its uncertainty (m and σ_m), the proper decay time and its uncertainty (ct and σ_{ct}), the flavor tag with its predicted dilution for both OST and SSKT (ξ and $\mathcal{D}$), and finally the three transversity angles $\Omega = (\cos \Psi, \cos \Theta, \Phi)$. Considering f_s the signal fraction in the sample, the entire PDF for an event is symbolically expressed as

$$\mathrm{PDF}(x_c|\vartheta) = f_s P_s(m|\sigma_m) P_s(ct, \Omega, |\sigma_{ct}, \xi, \mathcal{D}) P_s(\sigma_{ct}) P_s(\xi) P_s(\mathcal{D})$$
$$+ (1 - f_s) P_b(m) P_b(ct|\sigma_{ct}) P_b(\sigma_{ct}) P_b(\Omega) P_b(\xi) P_b(\mathcal{D}). \qquad (6.2)$$

We treat the background distributions (P_b) as fully separable in all the observable subspaces except for the decay-length PDF ($P_b(ct)$), which depends on decay-time resolution. For the signal, the full probability density function (P_s) is more complex. The full PDF construction is detailed in the following sections.

6.2.1 $J/\psi K K$ Mass PDF

The signal mass distribution $P_s(m|\sigma_m)$ is modeled by a Gaussian function with central value M, fixed to the known value of $5.3667 \pm 0.0004\,\mathrm{GeV}/c^2$ [3], and width equal to the candidate-specific mass resolution σ_m determined by the kinematic fit to the tracks scaled using a factor s_m to account for a common misestimation of mass uncertainties. The background mass model $P_b(m)$ is empirically chosen to be a straight line function, which models appropriately the dominant combinatorial background. Other sources of background such as $B^0 \rightarrow J/\psi K\pi$ are considered in the systematic uncertainties (Chap. 7). We do not include explicitly the PDF of σ_m in the fit because it has similar distribution for signal and background, as shown in Fig. 6.1. The effect of small residual differences is included in an alternative fit, described in Sect. 7.1.

Fig. 6.1 Mass uncertainty distributions for sideband-subtracted signal (*red*) and background (*blue*)

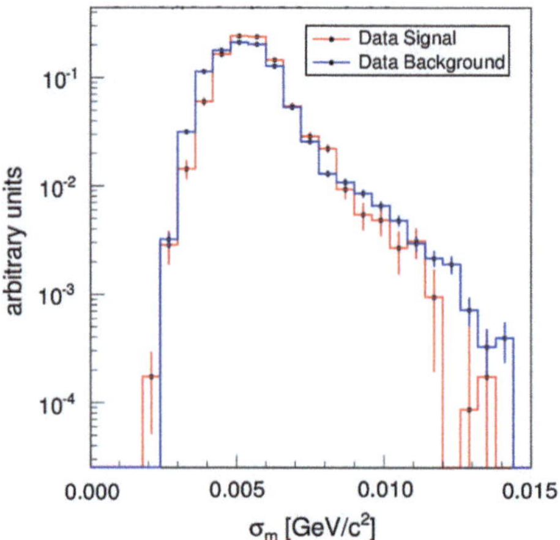

6.2.2 PDF of Angles and Decay Time for Signal

The PDF describing the distribution of transversity angles and ct of signal candidates is not separable and it is modeled by the differential decay rate as function of the transversity angles and proper decay time, $d^4\Lambda/(d^3\Omega\,dt)$.

Sculpting of the angular distributions caused by non-hermeticity of the detector and biases induced by the selection criteria are included along with the resolution on the measured ct of the candidate, σ_{ct}. The angular acceptance is observed to be independent of ct and is modeled by a multiplicative term, $\varepsilon(\Omega)$. Such a term is parametrized by an expansion in spherical harmonics and Legendre polynomial (Sect. 5.1).

The ct resolution is used to smear all terms describing the time evolution in the differential decay rate. Therefore, the exponential functions describing the decay and the oscillating functions for the mixing probability are convolved with the ct resolution function. This is empirically parametrized with a sum of two Gaussian functions whose parameters are extracted from the fit of the ct distribution of prompt background. Once the analytical form of the resolution function is given, the smeared terms, properly normalized, are used to replace the time evolution functions of the decay rate,

$$P(ct,\Omega|\sigma_{ct}) = \varepsilon(\Omega)d^4\Lambda/(d^3\Omega\,dt) \otimes \mathcal{R}(ct|\sigma_{ct}) \tag{6.3}$$

The analytical form of the differential decay rate $d^4\Lambda/(d^3\Omega dt)$ for $B_s^0 \to J/\psi\phi$ is found in Ref. [4], although we used a more compact formalism [5]. We included a potential S-wave contamination from decays in which the kaon pairs originate from decays of the scalar $f_0(980)$ resonance or are not resonant (Sect. 2.4.1).

6.2.2.1 Production Flavor PDFs

The decay rate for a meson initially produced as a $\bar{B}_s^0$ meson differs from the rate of an initial B_s^0 meson because of the sign flip in the oscillating terms in the decay rate. We exploit this difference for a more precise estimation of the mixing phase by introducing flavor tag information. In Eq. 6.2, we decompose the PDF in various components using conditional-probability, $P(\xi)$, $P(D)$, and $P(\Omega, ct|\sigma_{ct}, \mathcal{D}_i, \xi_i)$. The index in the last term labels the two tagging algorithms used in the analysis. The PDF that accounts for the tag decision takes into account the efficiencies ε_i of the two taggers (Sect. 5.3),

$$P(\xi) \equiv P(\xi_1)P(\xi_2) = \sum_{j=1}^{2}\varepsilon_j\delta\left(|\xi_j|-1\right) + \left(1-\sum_{j=1}^{2}\varepsilon_j\right)\delta\left(\sum_{j=1}^{2}\xi-0\right). \tag{6.4}$$

The PDF of the dilution ($P(\mathcal{D}_i)$) is modeled with an histogram from data. Independent histograms are used for signal and background, and for different tagging

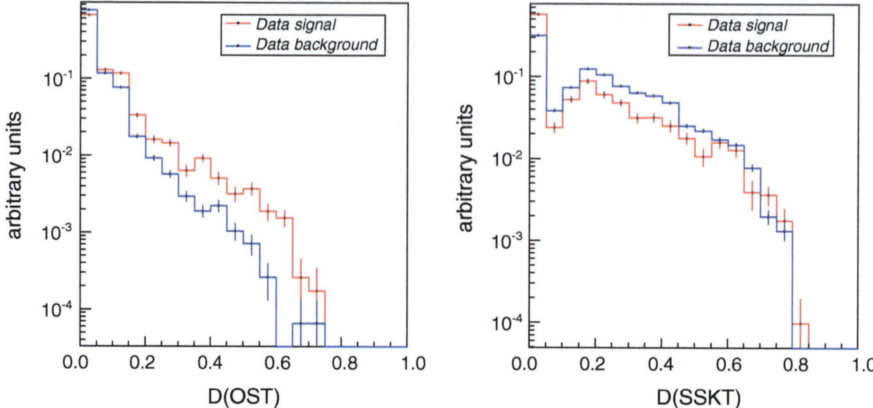

Fig. 6.2 Dilution histograms for OST (*left*) and for SSKT (*right*). Comparison between sidebands and sidebands-subtracted data

algorithms. The signal histograms are derived from background-subtracted data; the background dilution histograms are derived from mass-sidebands data. These distributions are reported in Fig. 6.2 for OST (left) and SSKT (right).

Tagging dilutions (both OST and SSKT) are multiplied by a scale factor s_i to account for mismodeling. These scale factors are separately extracted from a dedicated calibration of both taggers (Sect. 5.3). In the likelihood, each scale factor is free to float within a Gaussian constraint corresponding to the calibrated values.

Finally, the complete angular and decay-time PDF is

$$P(\Omega, ct|\sigma_{ct}, \mathcal{D}_1, \mathcal{D}_2, \xi_1, \xi_2) = \frac{1 + \xi_1 s_1 \mathcal{D}_1}{1 + |\xi_1|} \frac{1 + \xi_2 s_2 \mathcal{D}_2}{1 + |\xi_2|} P(\Omega, ct|\sigma_{ct})$$
$$+ \frac{1 - \xi_1 s_1 \mathcal{D}_1}{1 + |\xi_1|} \frac{1 - \xi_2 s_2 \mathcal{D}_2}{1 + |\xi_2|} \bar{P}(\Omega, ct|\sigma_{ct}). \quad (6.5)$$

Because we use SSKT only in the first 5.2 fb⁻¹ of data, the likelihood is factorized as follows:

$$\mathcal{L} = \mathcal{L}_{\text{OST+SSKT}} \times \mathcal{L}_{\text{OST}}, \quad (6.6)$$

where $\mathcal{L}_{\text{OST+SSKT}}$ refers to the likelihood of the first half of the sample, whose PDFs contains both OST and SSKT observables and parameters, while $\mathcal{L}_{\text{OST}}$ refers to the second part of the sample, where only the OST information is exploited. In the likelihood, $\mathcal{L}_{\text{OST}}$ has four fewer fit parameters than $\mathcal{L}_{\text{OST+SSKT}}$: the SSKT tagging efficiency for signal and background ($\varepsilon_s(\text{SSKT})$ and $\varepsilon_b(\text{SSKT})$), the SSKT background tag asymmetry ($A^+(\text{SSKT})$) and the SSKT dilution scale factor $S_D(\text{SSKT})$. Moreover, the SSKT dilution PDFs are not included in $\mathcal{L}_{\text{OST}}$, while the tag decision PDF Eq. (6.4) and the angular and decay-time PDF Eq. (6.5) only include the index for the OST case.

6.2.3 Background Lifetime PDF

Building on precise lifetime measurements performed at CDF in similar final states [6] three components are identified in the ct distribution of background events:

- a prompt narrow structure, which account for the majority of the combinatorial background events, expected to have no significant lifetime;
- two positive exponentials used to describe background that appears as being longer lived;
- a negative exponential for those background events that mimic a negative decay length in the vertex reconstruction.

The prompt peak of the background allows to determine the core of resolution function in ct. The resolution function is modeled using two Gaussian distributions,

$$\mathcal{R}(ct|\sigma_{ct}) = f_1 G_1(ct, s_{ct1}\sigma_{ct}) + (1 - f_1)G_2(ct, s_{ct2}\sigma_{ct}), \tag{6.7}$$

where f_1 is the fraction of the first Gaussian distribution, and s_{ct1} and s_{ct2} are scale factors of candidate specific ct-uncertainty, σ_{ct}, that are free to float in the fit.

6.2.4 Background Angular PDF

The background PDF of the transversity angles is parametrized empirically from sidebands of the B_s^0 mass distribution. Each transversity angle distribution is observed to be independent of the others with good approximation, $P(\Omega) = P(\cos\Theta)P(\Phi)P(\cos\Psi)$. The background angular distributions are also found to be independent of ct (Fig. 6.3). A satisfactory model consists in the following empirical functions: $f(\cos\Theta) \propto 1 - a\cos^2(\Theta)$, $f(\Phi) \propto 1 + b\cos(2\Phi)$ (where a and b are fit parameters), and $f(\cos(\Psi)) = \text{const}$. This model is simpler than the model used in the previous analysis [7] but equally accurate.

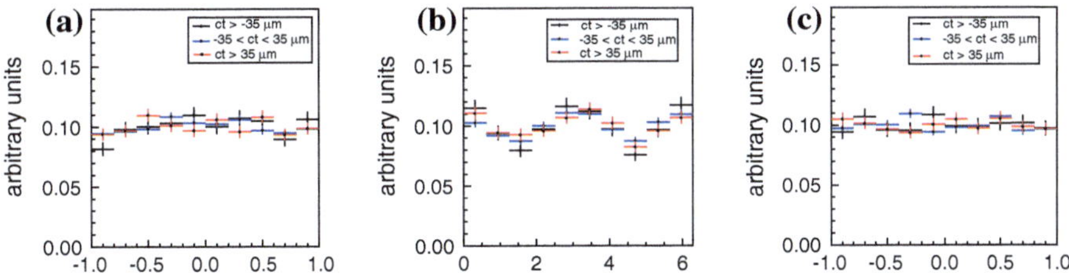

Fig. 6.3 Angular distribution of $\cos\Theta$ (**a**), Φ (**b**), and $\cos\Psi$ (**c**), for events restricted to various intervals of ct

6.2.5 Lifetime Uncertainty PDFs

The empirical PDF for the decay-time uncertainty uses Gamma functions as follows:

$$P(\sigma_{ct}) = f_P \frac{(\sigma_{ct})^{a_1} e^{\frac{\sigma_{ct}}{b_1}}}{(b_1)^{a_1+1}\Gamma(a_1+1)} + (1 - f_P)\frac{(\sigma_{ct})^{a_2} e^{\frac{\sigma_{ct}}{b_2}}}{(b_2)^{a_2+1}\Gamma(a_2+1)}, \tag{6.8}$$

where the a_1 b_1 a_2, and b_2 define the mean and the width of the first and the second distribution respectively, and f_P defines the fraction of the first distribution. Both the background and the signal PDF have the form of Eq. (6.8) with different parameters. These parameters are found with a preliminary lifetime-only fit on data. The parameters values are then used as input in the full likelihood used for the complete analysis.

6.2.6 Summary of the Fit Variables

The parameters of the maximum likelihood fit are listed in Table 6.1. Those parameters used to model the lifetime uncertainty PDF are not listed, since they are not floating, but they are determined with a previous lifetime-only fit. For convenience in writing the likelihood, a change of variables is applied to the parameters for the polarization amplitudes A_i. Following the formalism in [5], the polarization amplitudes at $t = 0$ are derived from the estimation of the time-integrated rate, $|a_i|^2$, to each of the polarization states. Exploiting the unitarity relation $\sum_{i=0,||,\perp} |a_i|^2 = 1$, we can fit the polarization rates using the fraction $\alpha_{||} = |a_{||}|^2/(1 - |a_\perp|^2)$ and $\alpha_\perp = |a_\perp|^2$ which relate to the amplitudes $|A_i|^2$ as follow:

$$|A_\perp|^2 = \frac{\alpha_\perp y}{1 + (y - 1)\alpha_\perp}, \tag{6.9}$$

$$|A_{||}|^2 = \frac{(1 - \alpha_\perp)\alpha_{||}}{1 + (y - 1)\alpha_\perp}, \tag{6.10}$$

$$|A_0|^2 = \frac{(1 - \alpha_\perp)(1 - \alpha_{||})}{1 + (y - 1)\alpha_\perp}, \tag{6.11}$$

with $y \equiv (1 + z)/(1 - z)$ and $z \equiv \cos 2\beta_s \Delta\Gamma_s/2\Gamma_s$.

6.3 Fit Validation and Checks

We perform several tests of the fit to validate the correctness of its implementation, to test the statistical limits of our sensitivity to the parameters of interest, and to investigate the likelihood properties under various circumstances. The procedure is

Table 6.1 Overview of fit parameters

Parameter	Description
$\beta_s^{J/\psi\phi}$	CP-violating phase of $B_s^0 - \bar{B}_s^0$ mixing amplitude
$\Delta\Gamma_s$	Lifetime difference $\Gamma_L - \Gamma_H$
$\alpha_\perp$	Time-integrated rate of polarization state $|a_\perp|^2$
$\alpha_\parallel$	Fraction of the time-integrated rate of polarization states $|a_\parallel|^2/(1 - |a_\perp|^2)$
$\delta_\perp$	$Arg(A_\perp A_0^\star)$
$\delta_\parallel$	$Arg(A_\parallel A_0^\star)$
$c\tau$	B_s^0 mean lifetime
$|A_S|^2$	Fraction of S-wave K^+K^- component in the signal
δ_S	Strong phase of S-wave amplitude
Δm_s	$B_s^0 - \bar{B}_s^0$ mixing frequency
f_s	Signal fraction
s_m	Mass uncertainty scale factor
p_1	Mass background slope
s_{ct_1}	Lifetime uncertainty scale factor 1
s_{ct_2}	Lifetime uncertainty scale factor 2
$f_{\mathcal{R}}$	Relative fraction of Gaussians of the decay-length resolution
f_{pr}	Fraction of prompt background
f_{Λ_n}	Fraction of background with Λ_n lifetime
f_{Λ_1}	Fraction of background with Λ_1 lifetime
Λ_n	Inverse of lifetime of a background component
Λ_1	Inverse of lifetime of a background component
Λ_2	Inverse of lifetime of a background component
a	Parameter in background fit to Φ
b	Parameter in background fit to $\cos\Theta$
$S_D(\mathrm{OST})$	OST dilution scale factor
$S_D(\mathrm{SSKT})$	SSKT dilution scale factor
$\varepsilon_b(\mathrm{OST})$	OST tagging efficiency for background
$\varepsilon_b(\mathrm{SSKT})$	SSKT tagging efficiency for background
$A(\mathrm{OST})$	OST background tag asymmetry
$A(\mathrm{SSKT})$	SSKT background tag asymmetry
$\varepsilon_s(\mathrm{OST})$	OST tagging efficiency for signal
$\varepsilon_s(\mathrm{SSKT})$	SSKT tagging efficiency for signal

as follows: assuming a given set of true values for the fit parameters, ϑ_{input}, we generate a large set of pseudoexperiments, randomly sampling the probability density function in each variable subspace to assign event variables. Each pseudoexperiment represents a statistically independent sample of events. We fit each of these pseudoexperiments as if they were experimental data. For each floating fit parameter ϑ, the pull distribution P is defined as

$$P = \frac{\vartheta_{fit} - \vartheta_{input}}{\sigma_\vartheta}, \tag{6.12}$$

where ϑ_{fit} and σ_ϑ are the ML estimate of the parameter under study and its uncertainty, respectively. We expect P to follow a Gaussian distribution centered at zero with unit variance for any set of input true values ϑ_{input}. Any deviation from Gaussian-distributed pulls indicates potential features of the likelihood that may need to be accounted for in the measurement. Examples of such occurrences are when a bound is imposed on the parameter range or when the actual dimensionality of the likelihood depends on the true value of some parameters. We generate 1,000 pseudoexperiments, each with similar event yield as in data. In the following section we report an overview of the most relevant of these tests in two different fit configurations: a fit in which the mixing phase is constrained to the SM value (*CP*-conserving fit) and a fit in which the phase is allowed to float (*CP*-violating fit).

6.3.1 CP-Conserving Fit

We first inspect the pull distribution of the parameters of highest physics interest: $\Delta\Gamma$, $c\tau$, $\alpha_\|$, $\alpha_\perp$, and the strong phases $\delta_\perp$ and $\delta_\|$ (Fig. 6.4). The pulls are regular for all quantities but $\delta_\|$ (Fig. 6.4f). The $\delta_\|$ pull shows a non-Gaussian behavior, which makes challenging to quote a reliable point estimate. The likelihood for the $\delta_\|$ parameter has a symmetry for reflection around $\delta_\| = \pi$; for estimated values close to the symmetric point π, the minimization process may not unambiguously differentiate the two symmetric, close minima and tends to return the boundary value, π, as estimate. Hence, we chose not to include a point estimate of $\delta_\|$ among the measurement's results.

The pull of $\alpha_\perp$ shows a bias of approximately 0.2 standard deviations. Also $\alpha_\|$ shows a bias of about 0.1 standard deviations. After verifying that the entity of these biases is moderate and independent of the true values of parameters, we decide to add such biases as fit-related systematic uncertainties.

The S-wave amplitude parameter A_{SW} is left free to float in the fit, but within a $[0,1]$ boundary to prevent A_{SW} from assuming unphysical negative values. These are not allowed in the likelihood, where the S_{wave} amplitude enters under a square root. The pull distribution for the S-wave amplitude A_{SW} and its phase δ_s with respect to the P-wave is shown in Fig. 6.4g and h. We observe a deviation from the Gaussian distribution of the A_{SW} pull, while the δ_s pull distribution appears more regular, even if not properly Gaussian. This is attributed to the choice of an A_{SW} value in the generation of the pseudoexperiments that is very close to the boundary $(1.8 \pm 2.3)\,\%$. That value is obtained from the fit to data in the previous CDF measurement [7]. A closer inspection to the A_{SW} estimator is given in Fig. 6.5a. For estimated values close to the boundary, the minimization tends to return the boundary value as estimate. Figure 6.5a shows indeed that more than half of the fits find an A_{SW} value that is either zero or in the interval $[0.0 - 0.02]$ while the residual in Fig. 6.5b clearly indicates

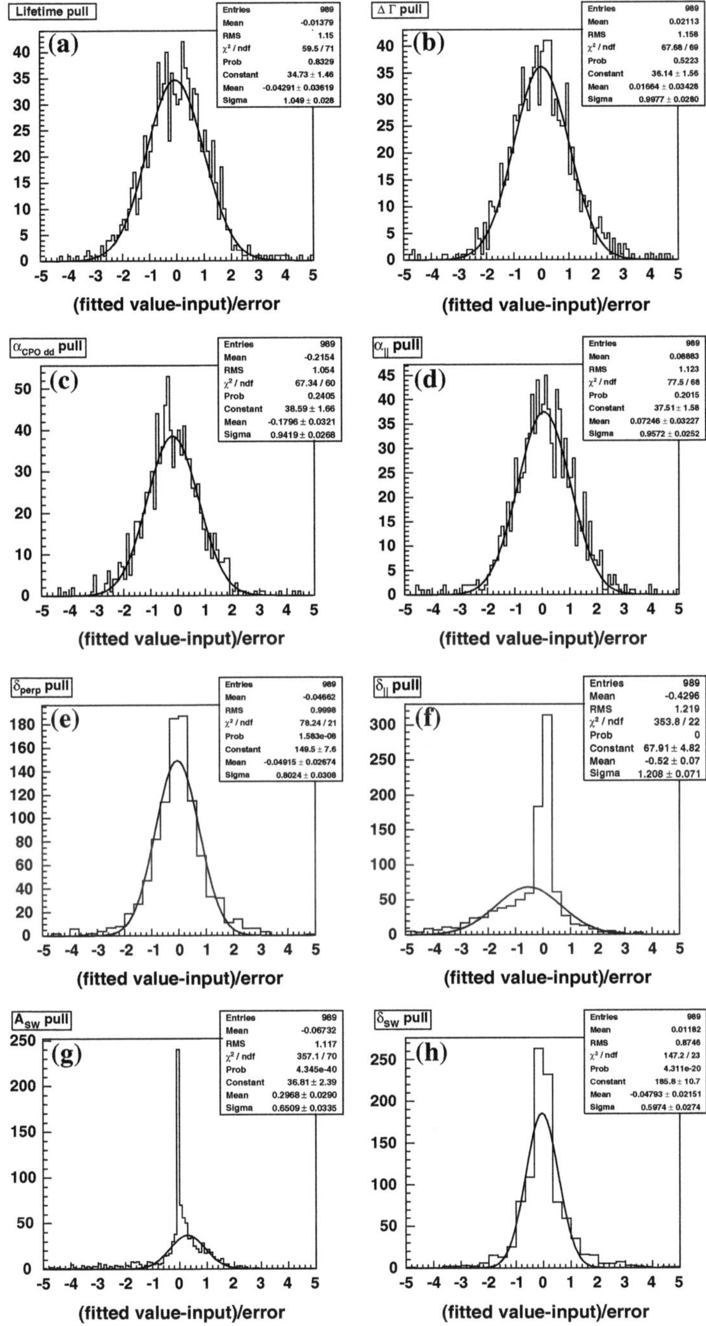

Fig. 6.4 Pull distributions for the main physics parameters of the *CP*-conserving fit

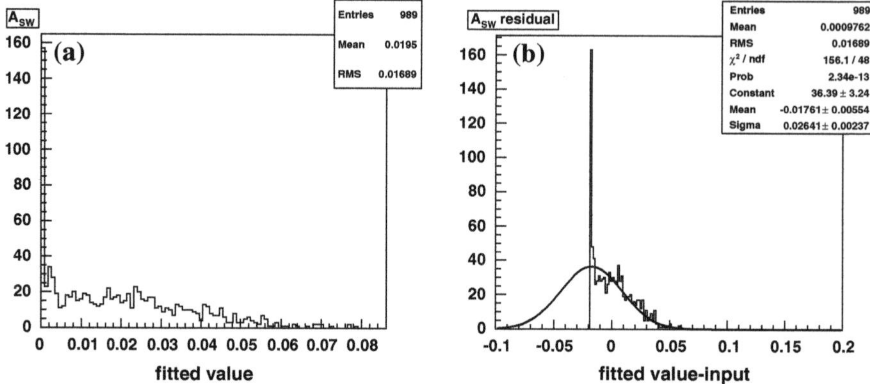

Fig. 6.5 Maximum likelihood estimate (**a**) and residual (**b**) distributions for the S_{wave} amplitude A_{SW} in pseudoexperiments generated with $A_{SW} = 1.8\,\%$, as obtained from previous CDF measurements [7]

a biased estimate. Hence, we conclude that the boundary at $A_{SW} = 0$ induces the values found for A_{SW} and deforms the pull distribution (Fig. 6.4g).

In addition, the fit occasionally shows numerical problems, because the minimizer finds an absolute minimum at the limit of the A_{SW} domain (see Ref. [8]). We fix those cases by initiating another minimization from a local minimum of the likelihood, moving the starting point of A_{SW}. To support the hypothesis that the boundary is biasing the A_{SW} estimator, another set of pseudoexperiments is generated, with generation value of $A_{SW} = 25\,\%$, far away from the boundary. We expect a Gaussian distribution centered at 0.25 for the A_{SW} estimates. Figure 6.6 shows the resulting distribution, in agreement with the expectations.

The mean fit uncertainty for the physics parameters in the simulated experiments is comparable with the uncertainty observed in data. This comparison is reported in Table 6.2, together with the parameters describing the pull distributions (mean and width). The maximum likelihood values are reliable, and the observed uncertainties agree with expectation from pseudoexperiments, except for the uncertainty on $\delta_\perp$, which is larger than estimated in data. In a large fraction of pseudoexperiments, the fit may converge to a minimum different from the generation value, due to the proximity of the two minima. In such cases the fit uncertainty increases. However, the difference between the mean uncertainty as extracted from pseudoexperiments and the fit uncertainty in data is compatible with the systematic uncertainty (Chap. 7) associated to $\delta_\perp$ from pull studies, suggesting that this feature should not be an issue for the final estimation of this parameter.

6.3.2 Independence from Nuisance Parameters

To test the independence of the ML estimates from true values of nuisance parameters, we repeat the pull study by generating 2,000 pseudoexperiments in which the true

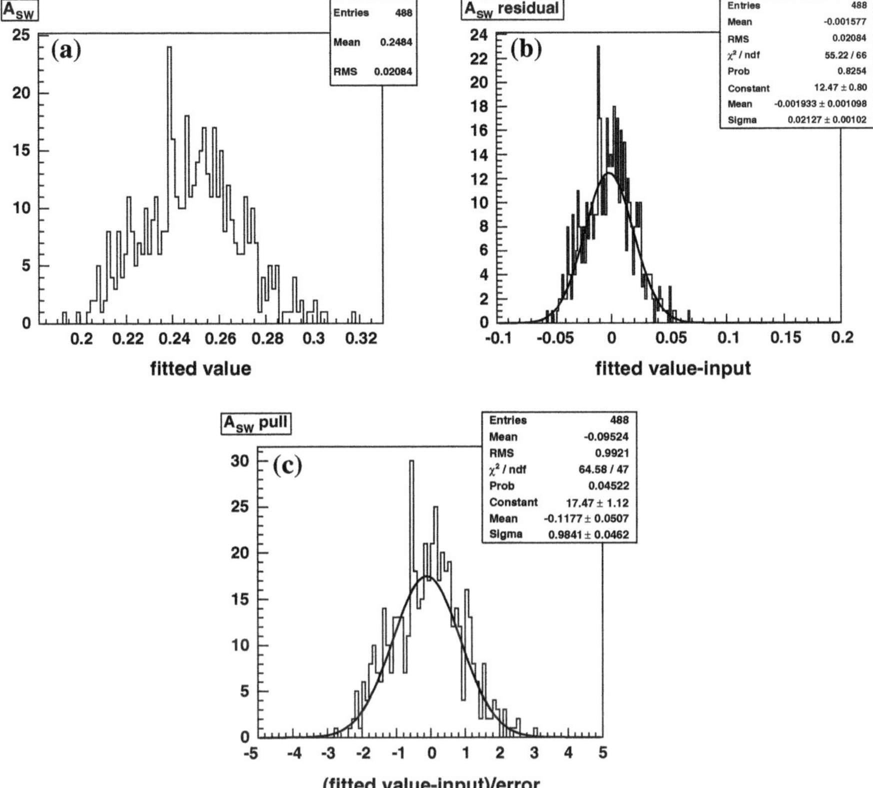

Fig. 6.6 Maximum likelihood estimate (**a**), residual (**b**) and pull (**c**) distributions for the S_{wave} amplitude A_{SW} in pseudoexperiments generated with $A_{SW} = 25\,\%$

Table 6.2 Mean and σ of the pull distribution

Parameter	Pull mean	Pull σ	Mean uncertainty	Fit uncertainty in the data
$c\tau$	-0.043 ± 0.036	1.049 ± 0.028	0.00061	0.00062
$\Delta\Gamma_s$	0.017 ± 0.034	0.998 ± 0.028	0.028	0.029
$\alpha_\perp$	-0.180 ± 0.032	0.942 ± 0.028	0.012	0.012
$\alpha_\parallel$	0.072 ± 0.032	0.957 ± 0.027	0.013	0.012
$\delta_\perp$	-0.049 ± 0.027	0.802 ± 0.030	0.58	0.612

Variable mean uncertainty and in the last column the fitted parameter uncertainty

values of nuisance parameters are sampled randomly. The range of sampled values is chosen to be centered at the ML estimate in data and extends to ±5 times the estimated uncertainties. A result summary of this test is reported in Fig. 6.7. No significant dependence on the true values of nuisance parameters is observed.

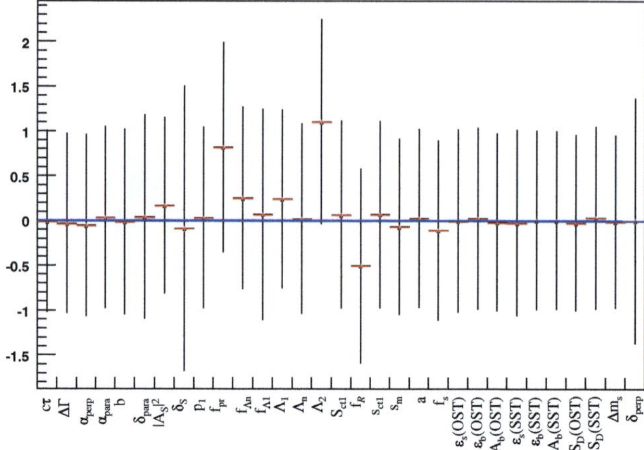

Fig. 6.7 Summary of the pull distributions mean and width for all parameters of the likelihood in which β_s is fixed. True values of input parameters are chosen randomly from their physical domain. The *red* markers and the *black lines* are the means and the widths of the observed pull distributions, respectively. The *blue line* is drawn as reference (color figure online)

6.3.3 CP-Violating Fit

We validated the fit in which the mixing phase phase is free to vary using pseudoexperiments generated with various true values of $\beta_s^{J/\psi\phi}$ and values of nuisance parameters as used for the *CP*-conserving fit studies. Figure 6.8 shows the pull distribution for the main physics parameters. Significant bias is observed for the $\Delta\Gamma$ estimate. Contrary to previous CDF measurement, Fig. 6.9. shows that no significant bias is affecting $\beta_s^{J/\psi\phi}$, probably owing to the larger data sample size now available. Similar considerations as for the *CP*-conserving case hold for the $\delta_\parallel$, A_{SW}, and δ_s estimates.

In Table 6.3 we report a comparison between the mean uncertainties on a given parameter observed in the pseudoexperiments and the uncertainties obtained by fitting data. Again, the considerations made in the *CP*-conserving case are still applicable here.

6.4 Final Considerations on the Likelihood Fit

The ML fit yields unbiased or minimally biased estimates of the key physics parameters with $\beta_s^{J/\psi\phi}$ fixed to its expected value in the SM (0.02). The fit parameters for which point estimates are statistically reliable are the proper decay length, $c\tau$, the decay-width difference, $\Delta\Gamma_s$, the polarization fractions $\alpha_\perp$ and $\alpha_\parallel$, and the strong phase $\delta_\perp$.

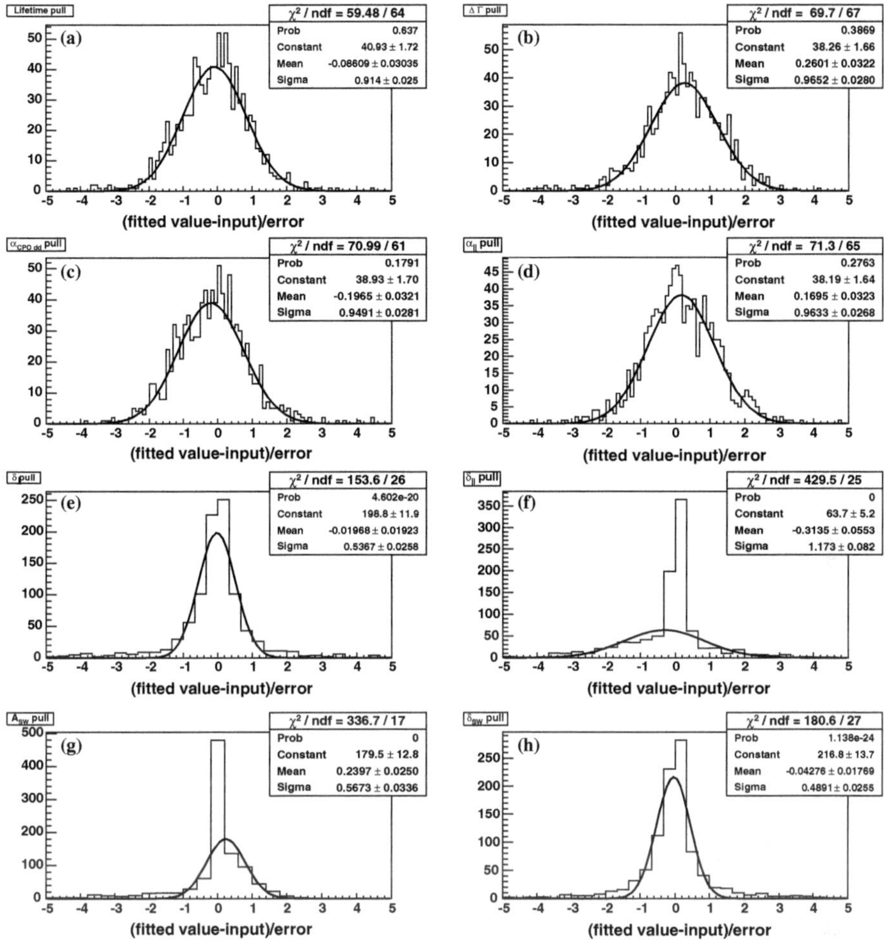

Fig. 6.8   Pull distributions for the main physics parameters for the CP-violating fit

When the mixing phase $\beta_s^{J/\psi\phi}$ is left free to vary, the fit shows significant biases on $\Delta\Gamma_s$. In the previous iteration of the analysis the origin of this has been carefully tracked down to [7]

1. a combination of the complications due to likelihood symmetries, which introduce multiple, equivalent solutions,
2. the fact that the sensitivity to some parameters mathematically depends on the estimated values of others,
3. and the current sample size being still insufficient to approximate the asymptotic regime for the given likelihood.

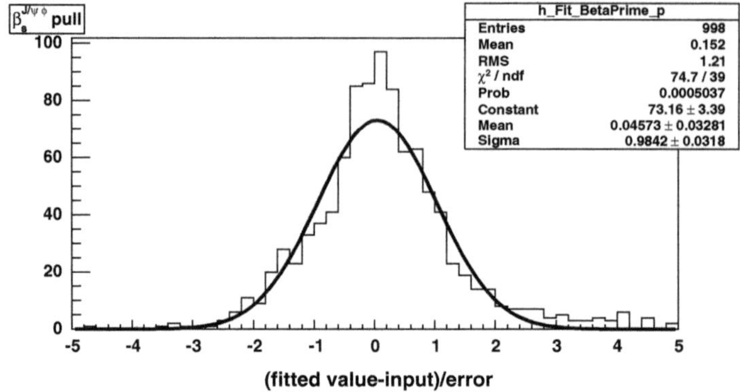

Fig. 6.9 $\beta_s^{J/\psi\phi}$ pull distribution

Table 6.3 Mean and σ of the pull distribution corresponding to the 10 fb^{-1} CP-violating fit

Parameter	Pull mean	Pull σ	Mean uncertainty	Fit uncertainty in the data
$\beta_s^{J/\psi\phi}$	0.046 ±0.033	0.984±0.032	0.197	0.123
$c\tau$	−0.086±0.030	0.914±0.025	0.00069	0.00058
$\Delta\Gamma_s$	0.026±0.032	0.965±0.028	0.038	0.028
$\alpha_\perp$	−0.197±0.032	0.949±0.028	0.012	0.011
$\alpha_\parallel$	0.170±0.032	0.963±0.027	0.013	0.011
$\delta_\perp$	−0.020±0.019	0.537±0.026	1.701	0.739

Hence, we decided not to quote point estimates from the CP-violating fit and report results in the form of confidence intervals by resorting to the Neyman construction of a fully frequentist confidence region [9].

References

1. F. James et al., Minuit, function minimization and error analysis. Minuit Reference Manual (2007)
2. R. Brun, F. Rademakers, S. Panacek, ROOT, an object oriented data analysis framework, 11 (2000)
3. J. Beringer et al., (Particle Data Group), Review of particle physics (RPP). Phys. Rev. **D86**, 010001 (2012)
4. I. Dunietz et al., In pursuit of new physics with B_s decays. Phys. Rev. D. **63**, 114015 (2001), arXiv:hep-ph/0012219 (hep-ph)
5. F. Azfar et al., Formulae for the analysis of the flavor-tagged decay $B_s^0 \to$ Jpsi phi. JHEP **1011**, 158 (2010), arXiv:1008.4283 (hep-ph)
6. T. Aaltonen et al., (CDF Collaboration), Measurement of b hadron lifetimes in exclusive decays containing a $B_s^0 \to J/\psi\phi$ in pp collisions at $\sqrt{s} = 1.96$ TeV. Phys. Rev. Lett. **106**, 121804 (2011)

7. T. Aaltonen et al. (CDF Collaboration), Measurement of the *CP*-violating phase $\beta_s^{B_s^0 \to J/\psi\phi\phi}$ in $B_s^0 \to J/\psi\phi$ decays with the CDF II detector. Phys. Rev. D **85**, 072002 (2012), arXiv:1112.1726 (hep-ex)
8. F. James, CERN program library long writeup D506
9. J. Neyman, Outline of a theory of statistical estimation based on the classical theory of probability

Chapter 7
Systematic Uncertainties

In this chapter we discuss the systematic uncertainties considered in the measurement.

7.1 General Approach

Several assumptions are made in the description of the likelihood, and additional experimental and physics effects are not fully accounted for. We consider many of these as sources of potential systematic uncertainty. Systematic uncertainties are assigned to include effects from potential mismodeling in the fit, or physical effects that are not well known or fully incorporated into the model. For each considered systematic effect, two sets of about 500 pseudoexperiments each are generated, one by randomly drawing events sampled from the likelihood corresponding to the default fit model, the other using a fit model appropriately modified to reproduce each individual effect considered. The pseudoexperiments of the two sets are generated using the same seeds for the pseudorandom generator, in order to minimize the impact of statistical fluctuations. For each observable, we study the distribution of differences between the result of the fit on the default pseudoexperiments and the results on the modified pseudoexperiments. The systematic uncertainty associated to each observable is then taken to be the mean of the distribution of these differences

$$\sigma_{\text{syst}} = |\langle \vartheta_{\text{syst}}^{\text{fit}} - \vartheta_{\text{ref}}^{\text{fit}} \rangle|, \tag{7.1}$$

where the subscripts syst and ref stand for the pseudoexperiment generated according to the modified or default models, respectively. When the statistical uncertainty on the mean value of the observed difference exceeds the difference itself, we choose to consider the statistical uncertainty, rather than the central value, as systematic uncertainty. The B_s^0 lifetime, width-difference and transversity amplitudes values to be used in the simulation are taken from a previous CDF measurement [1]. In the following we list the considered systematic effects.

© Springer International Publishing Switzerland 2015

S. Leo, *CP Violation in $B_s^0 \to J/\psi\phi$ Decays*, Springer Theses,

DOI 10.1007/978-3-319-07929-5_7

7.1.1 Signal Angular Efficiency

One source of systematic uncertainty is the modeling of the angular efficiency of
the detector described in Sect. 5.1. The detector efficiency is modeled with a linear combination of Legendre polynomials and spherical harmonics. The expansion
coefficients of these functions are obtained by fitting a three-dimensional efficiency
distribution obtained using simulated events. This simulated sample is reweighted
to match the kinematic distributions observed in data. If the modeling or the p_T
reweighting is inaccurate, a systematic bias could be introduced. To probe the impact
of an extreme mismodeling, we generated pseudoexperiments using an unreweighted
simulated sample to derive the detector acceptance (Fig. 7.1) and fit them using the
default fit model. This source of systematic uncertainty represents the leading contribution to the systematic uncertainties on the squared magnitudes of the polarization
amplitudes.

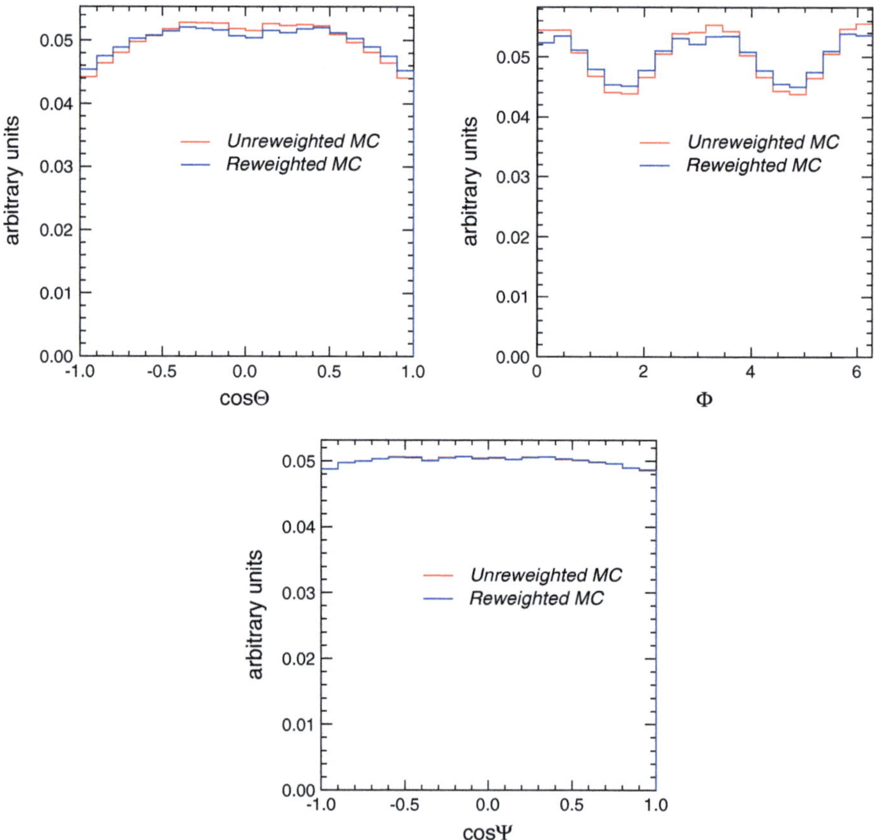

Fig. 7.1 Comparison between angular distributions of reweighed and non-reweighed simulation

7.1.2 Signal Mass Model

The B_s^0 mass distribution is fitted by default with a single Gaussian distribution. Possible mass-shape mismodelings could introduce biases. To test the size of a potential systematic effect, an alternative PDF for the signal mass is used, which consists of two Gaussian distributions, with same mean and independent widths. The relative sizes of the two areas and widths are left floating in a fit to data. Pseudoexperiments with a double-Gaussian signal mass model, described by the parameters extracted from the fit on data, are generated and fit with the default single Gaussian parameterization. The impact on the systematic uncertainties is moderate.

7.1.3 Background Mass Model

Similarly, the straight-line model used for the mass distribution of combinatorial background events could contribute a systematic uncertainty, if inadequate to describe the data. The systematic uncertainty associated with the background mass model is calculated by fitting the B_s^0 sideband mass distribution in data with a second-order polynomial function, and generating pseudoexperiments according to the resulting fit parameters. The pseudoexperiments are then fit with the default straight-line function. The impact of this source on systematic uncertainties is limited.

7.1.4 Lifetime Resolution Model

The lifetime measurement is crucially sensitive to the accuracy of the lifetime resolution model. In the central fit, the time-dependent component of the likelihood is convoluted with a two-Gaussian resolution function. To test the effect of a misparameterization of the resolution function, an alternative three-Gaussian resolution model is used. The parameters of the three-Gaussian model are used to generate a set of pseudoexperiments subsequently fitted with the default model. This systematic shift has a significant impact on the $\delta_\perp$ determination, with a smaller impact on the other parameters of interest.

7.1.5 Background Lifetime Fit Model

The modeling of the various components of the effective lifetime background (Sect. 6.2.3) can systematically affect the B_s^0 lifetime measurement. We include an alternative background-lifetime model using the histogram of the proper time of the events populating the signal mass sidebands (with ct resolution parameters fixed at the results of the default fit). The same histogram used to build the PDF for the alternative fit is used to generate pseudoexperiments that are then fitted with the default model. This has a relevant impact on several observables. It represents nearly 80 % of the $\Delta\Gamma_s$ total systematic uncertainty. An impact in excess of 40 % is also observed on the determination of the B_s^0 lifetime and $\delta_\perp$.

7.1.6 Angular Background Model and Correlations

A similar strategy is used to estimate the size of the contribution due to a mismodeling of the angular background distribution (Sect. 6.2.4). We use histograms of the transversity angles of events populating the mass data sidebands to generate pseudoexperiments that are fitted with the default parametrization (Sect. 6.2.4). Also, two-dimensional histograms of the angles versus σ_{ct} are used to probe potential biases from neglected correlations among these quantities. This results in a modest impact on the main observables.

7.1.7 Signal-to-Background Differences in Mass Uncertainty

In the central fit, we assume the mass resolution to be the same for signal and background events. The effect of any inaccuracy in this assumption is tested by using an alternative fit that includes the mass uncertainty distributions modeled by histograms restricted to B_s^0 sidebands for background events and sideband-subtracted signal-region for signal events, separately (Fig. 6.1). To generate the pseudoexperiments, we sample the background mass uncertainties from separate upper and lower sideband histograms. This accounts also for any bias from small correlations between σ_m and the reconstructed invariant mass, which are not included in the central fit. This affects significantly ($\approx$31 %) the lifetime total systematic uncertainty determination with a minor impact on other observables.

7.1.8 ct Uncertainty Model

To account for a possible misparametrization of the ct uncertainty distributions, an alternative fit exploiting the ct uncertainty distributions taken from data histograms is used rather than the model of Sect. 6.2.5. The PDFs used in the alternative fits are σ_{ct} histograms of B_s^0 sideband data for background events and sideband-subtracted data for signal events. To generate the pseudoexperiments, we sample the background uncertainties from separate higher- and lower-mass sideband histograms. This accounts for any effect caused by small observed correlations between σ_{ct} and the invariant mass. This is one of the dominant effect for the lifetime and width difference determinations, impacting their total systematic uncertainty by about 23 and 9 % respectively.

7.1.9 $B^0 \to J/\psi K\pi$ Cross-Feed

The $B^0 \to J/\psi K\pi$ decays misreconstructed as $B_s^0 \to J/\psi K K$ decays are not included in the default fit, though a fraction of these leaks into the B_s^0 signal region. We estimate the size of this contribution from data to be (7.99 ± 0.20) % of the B_s^0 signal (Appendix A). We generate pseudoexperiments according to this fraction and fit with the default model which does not account for this component. The effect from this contribution has limited impact on the final results.

7.1.10 SVX Alignment

A systematic uncertainty may arise from imperfect alignment of the silicon detector. A previous study on the effect of the limited knowledge of the CDF silicon detector alignment [2] concluded that estimate of the systematic uncertainty on the average decay length $c\tau$ in lifetime measurements is $\pm 2\,\mu$m. This study was done by fully reconstructing a number of exclusive $B \to J/\psi X$ decays. In both data and simulation under different silicon alignment assumptions, including correlated or independent shifts of $\pm 50\,\mu$m in all silicon detector components. The lifetime was fit and the largest deviation in results from the central values was taken as the systematic uncertainty on the lifetime due to the assumption of nominal silicon alignment.

The value of $\pm 2\,\mu$m systematic uncertainty on $c\tau(B_s^0)$ can also be used to assess secondary effects on the other parameters of interest. Due to correlations between the B_s^0 lifetime and the other physics parameters, it is reasonable to expect that an additional uncertainty on the lifetime measurement propagates into additional uncertainties in the measurements of other parameters. To quantify this contribution to the other parameters' uncertainties, pseudoexperiments in which the decay time in each event is randomly shifted by $\pm 2\,\mu$m are generated and fit with the default fit. The mean of the differences between fit results is used as systematic uncertainty.

7.1.11 Fit Biases

We assign a systematic uncertainty to parameters whose ML estimates show a bias, as studied in Sect. 6.3, which consists in the mean shift $\Delta_i = |\vartheta_i^{\text{fit}} - \vartheta_i^{\text{gen}}|$ where ϑ_i^{fit} is the estimate in the pseudoexperiments, and ϑ_i^{gen} is the value used in the generation of the pseudoexperiments, corresponding to the estimate in data.

Table 7.1 Systematic uncertainty summary

Source	$c\tau$ (μm)	$\Delta\Gamma$ (ps^{-1})	$|A_\parallel(0)|^2$	$|A_0(0)|^2$	$\delta_\perp$
Signal angular efficiency	0.29	0.0014	0.0134	0.0162	0.076
Mass signal model	0.17	0.0007	0.0006	0.0020	0.018
Mass background model	0.14	0.0006	0.0003	0.0002	0.034
ct Resolution	0.52	0.0010	0.0004	0.0002	0.066
ct Background	1.31	0.0057	0.0006	0.0012	0.064
Angular background	0.46	0.0037	0.0011	0.0022	0.009
Mass uncertainty	0.85	0.0006	0.0003	0.0002	0.036
ct Uncertainty	0.63	0.0006	0.0003	0.0002	0.038
$B_d \to J/\psi K^*$ cross-feed	0.42	0.0055	0.0009	0.0058	0.039
Vetex detector alignment	2.0	0.0004	0.0002	0.0001	0.034
Pull bias	0.2	0.0012	0.0021	0.0008	0.02
Total	2.8	0.009	0.014	0.018	0.15
Statistical	5.7	0.026	0.010	0.012	0.53

7.2 Total Systematic Uncertainty

Final systematic uncertainties obtained with the strategy described above are reported in Table 7.1. They are smaller than statistical uncertainties for some parameters such as $\Delta\Gamma_s$ but dominate the uncertainty on the polarization amplitudes.

A different approach is also used to verify the reliability of the estimation of systematic uncertainties. When possible, an alternative fit on data is performed whose likelihood function includes the potential systematic mismodelings, and the difference between the estimation of the observables with the alternative fit model and the values obtained with the default fit model is taken as systematic uncertainty. The purpose of these alternative fits is to provide an independent estimate of the size of the systematics uncertainties to cross check the estimate made by using pseudoexperiments. Very good agreement is found between the two approaches. Hence, for the final results we choose the one performed with pseudoexperiments since it is less affected from statistical fluctuations.

References

1. K. Anikeevs, *Measurement of the lifetimes of B_s^0 meson mass eigenstates*, Ph.D. thesis, Massachusetts Institute of Technology, available as fermilab-thesis-2004-12
2. S. Menzemer, *Track reconstruction in the silicon vertex detector of the CDF II experiment*, Ph.D. thesis, Karlsruhe University, 2003, available as FERMILAB-THESIS-2003-52, IEKP-KA-2003-04

Chapter 8
Results

In this chapter we present and discuss the results of the maximum likelihood fit to the time-dependent angular distributions of flavor tagged $B_s^0 \to J/\psi\phi$ decays. We first report results from the fit in which the CP-violating phase $\beta_s^{J/\psi\phi}$ is fixed to its SM value and then the results from the measurement of CP violation. Projections of the results of the likelihood fits in relevant distributions are also shown to ensure that the assumed models for the data distributions are accurate. The final results are compared with recent results from other experiments and theory predictions.

8.1 Lifetime, Decay-Width Difference, and Polarization Amplitudes

The fit with mixing phase fixed to the SM values is used to determine the lifetime, decay-width-difference and polarization amplitudes. Fit projections are in Figs. 8.1 and 8.2. The proper-decay time projection (Fig. 8.1) shows an overall agreement of background and signal distributions with the fit model. This is particularly evident in Fig. 8.1b, which shows projections for a signal-enriched sample, obtained by restricting the sample to events populating the signal region, $5.34 < m_{B_s^0} < 5.39\,\mathrm{GeV}/c^2$, and subtracting from them the distribution of events in the background-rich region (sideband), $(5.291 < m_{B_s^0} < 5.315\,\mathrm{GeV}/c^2) \cup (5.417 < m_{B_s^0} < 5.442\,\mathrm{GeV}/c^2)$. The chosen model shows a good ability to reproduce data distributions separately for signal and background (Fig. 8.1c) events. In this measurement, the significant fraction of prompt background (with vanishing lifetime) is exploited by the fit to precisely constrain the modeling of the resolution function in ct, yielding a precise determination of the scale factors for the decay-length uncertainty. The accuracy of the ct-resolution model is supported by the excellent fit agreement with data at negative values of ct, where the shape of the distribution is dominated by resolution effects. Fit projections for the transversity angles in the signal and

© Springer International Publishing Switzerland 2015

S. Leo, *CP Violation in $B_s^0 \to J/\psi\phi$ Decays*, Springer Theses,
DOI 10.1007/978-3-319-07929-5_8

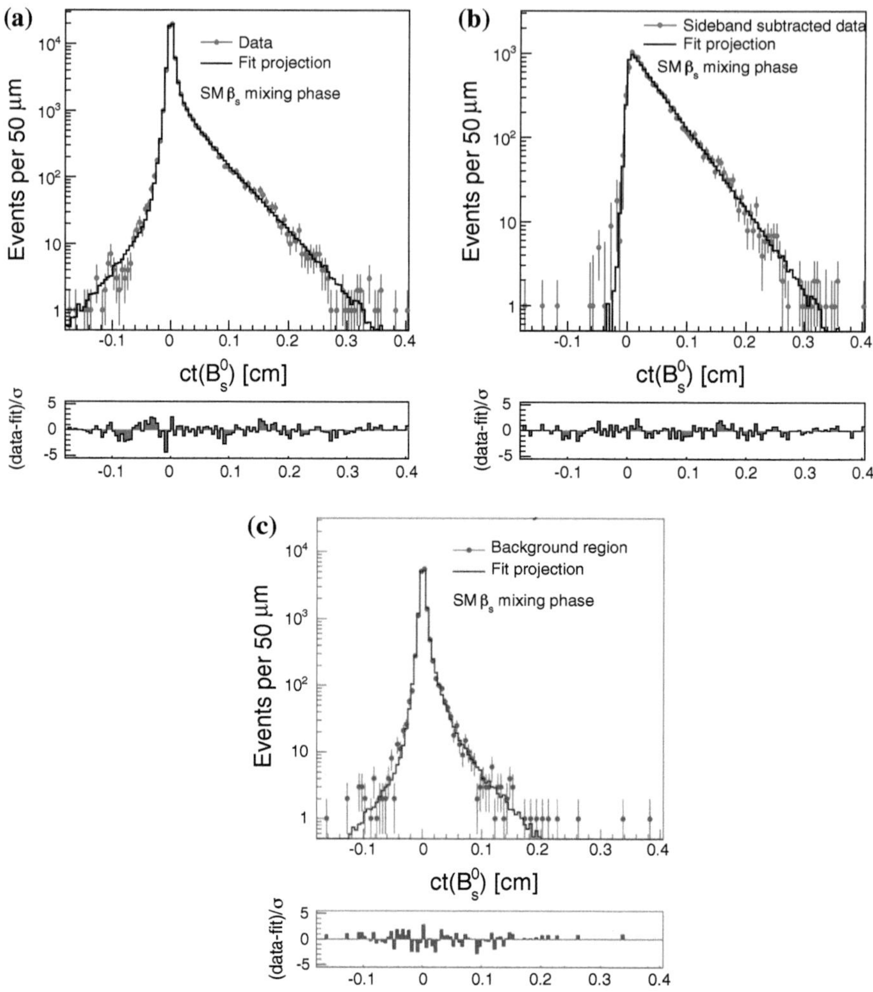

Fig. 8.1 Proper decay time projection for the SM fit in the complete fit range (**a**), and for events restricted to the $5.34 < m_{B_s^0} < 5.39\,\mathrm{GeV}/c^2$ region whose distribution is subtracted of sideband data distributions (**b**) and in the mass sideband region, with $(5.291 < m_{B_s^0} < 5.315\,\mathrm{GeV}/c^2)\ \cup\ (5.417 < m_{B_s^0} < 5.442\,\mathrm{GeV}/c^2)$ (**c**)

sideband regions are shown in Fig. 8.2. The good agreement validates the angular parameterization of both the signal and background distributions.

In Fig. 8.3 the scans of the likelihood for $\Delta\Gamma_s$, $\delta_\perp$ and $\delta_\parallel$ are shown. Each scan is obtained by calculating the value of the likelihood at various values of the corresponding parameter with all other parameters fixed at the values obtained in the global minimum of the likelihood. The scan is not a profile-likelihood (re-minimization of all parameters for each point), and provides only a qualitative idea of the shape of

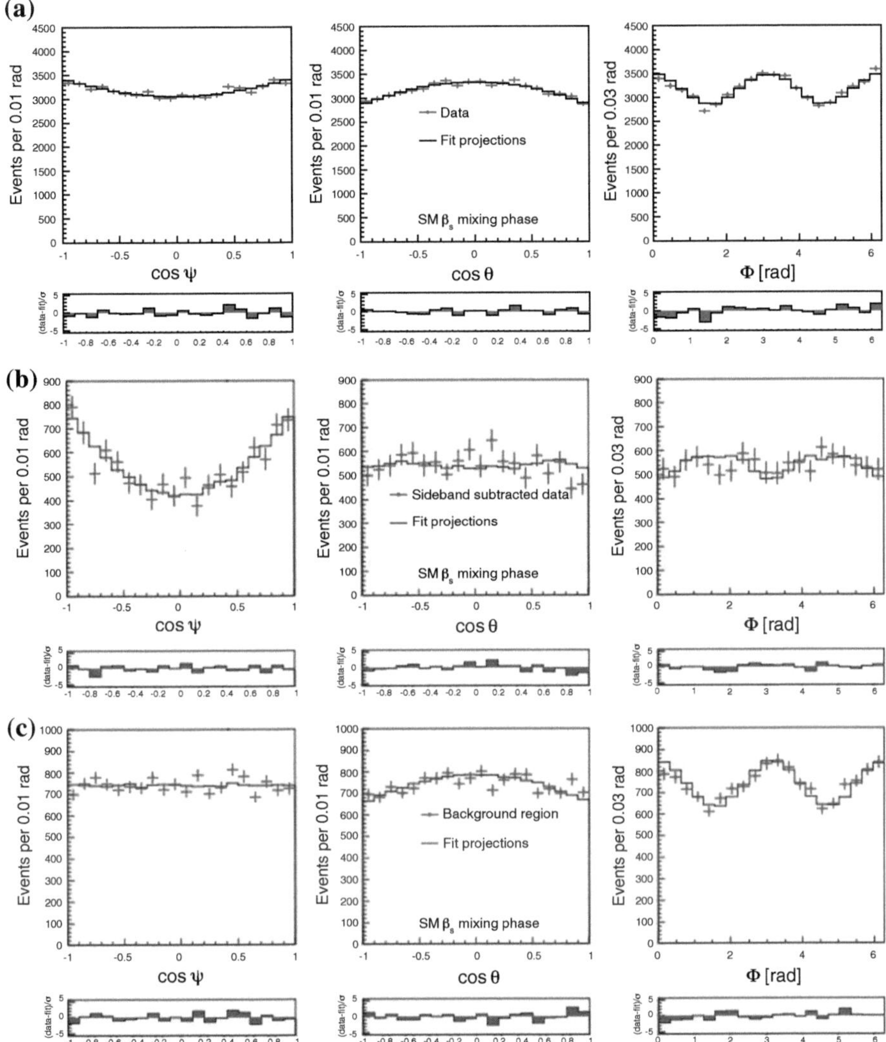

Fig. 8.2 Angular distributions, along with projection of the SM fit, for the complete fit range (**a**), for events restricted to the $5.34 < m_{B_s^0} < 5.39\,\mathrm{GeV}/c^2$ region whose distribution is subtracted of sideband data distributions (**b**) and in the mass sideband region, with $(5.291 < m_{B_s^0} < 5.315\,\mathrm{GeV}/c^2) \cup (5.417 < m_{B_s^0} < 5.442\,\mathrm{GeV}/c^2)$ (**c**)

the likelihood around the minimum of the parameter. The scans show a regular, parabolic shape for the former two parameters, and a non-parabolic shape for the latter, confirming previous findings (Chap. 6).

Table 8.1 reports the full fit results. Inspection of the technical fit parameters show that the background at $ct < 0$, which is due to misreconstructed events, exhibit similar effective lifetime as the short-lived the background at $ct > 0$, in agreement

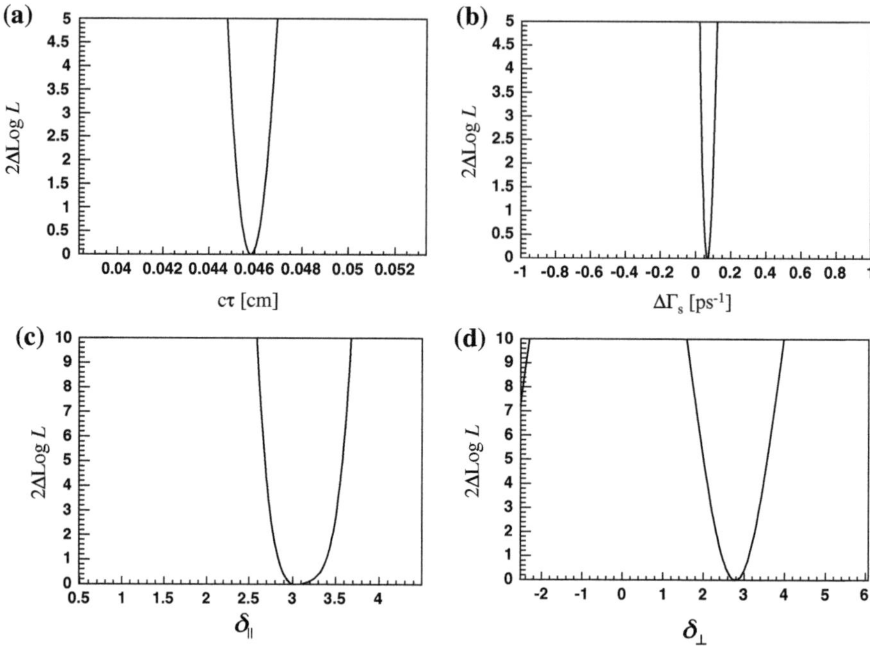

Fig. 8.3 Scan of the likelihood for ct (**a**), $\Delta\Gamma_s$ (**b**), $\delta_\parallel$ (**c**), and $\delta_\perp$ (**d**)

with the expectation that such misreconstruction effect is symmetric with respect to $ct = 0$. The efficiencies of either OST and SSKT taggers are in agreement with expectations from the calibrations (Sect. 5.3).

Since the estimate of $\delta_\parallel$ is biased, we choose not to report its point estimate. A similar consideration holds for the estimation of δ_s. The S-wave contribution in the mass region $1.009 < m_{KK} < 1.028$ GeV/c^2 as measured in data does not allow quoting a reliable estimation of the phase, being the observed fraction compatible with zero. A dedicated, alternative fit is discussed in Appendix A for a more precise measurement of the S-wave fraction and a consistency check of the default result.

Fit results referring to the main physical observables are reported in Table 8.2 along with correlations in Table 8.3. In Table 8.4 we also compare the current results with those from the previous CDF analysis [1], based on a subset of present data corresponding to $5.2\,\mathrm{fb}^{-1}$ of integrated luminosity. We observe very good consistency for all the results and observe a reduction in statistical uncertainty as expected from the increase in sample size.

Table 8.1 Results of the fit with $\beta_s^{J/\psi\phi}$ fixed to the SM value using the whole data set

Parameter	Description	Value
$\beta_s^{J/\psi\phi}$	CP-violating phase of B_s^0-$\bar{B}_s^0$ mixing amplitude	Fixed to 0.02
$\Delta\Gamma_s$	Lifetime difference $\Gamma_L - \Gamma_H$	0.068 ± 0.026 ps^{-1}
$\alpha_\perp$	Time-integrated rate of polarization state $|a_\perp|^2$	0.279 ± 0.010
$\alpha_\parallel$	Fraction of the time-integrated rate of polarization states $|a_\parallel|^2/(1 - |a_\perp|^2)$	0.309 ± 0.012
$\delta_\perp$	$Arg(A_\perp A_0^\star)$	2.79 ± 0.53
$\delta_\parallel$	$Arg(A_\parallel A_0^\star)$	3.09 ± 0.36
$c\tau_s$	B_s^0 mean lifetime	$458.2 \pm 5.8\,\mu$m
$|A_S|^2$	Fraction of S-wave K^+K^- component in the signal	0.008 ± 0.018
δ_S	Strong phase of S-wave amplitude	1.26 ± 0.75
Δm_s	B_s^0-$\bar{B}_s^0$ mixing frequency	17.74 ± 0.11 ps^{-1}
f_s	Signal fraction	0.1721 ± 0.0018
s_m	Mass uncertainty scale factor	1.727 ± 0.017
p_1	Mass background slope	-1.89 ± 0.42 (GeV/c^2)$^{-1}$
s_{ct_1}	Lifetime uncertainty scale factor 1	1.308 ± 0.012
s_{ct_2}	Lifetime uncertainty scale factor 2	3.34 ± 0.13
$f_\mathcal{R}$	Relative fraction of Gaussians of the decay-length resolution	0.852 ± 0.010
$f_{\rm pr}$	Fraction of prompt background	0.8839 ± 0.0039
f_{Λ_n}	Fraction of background with Λ_n lifetime	0.210 ± 0.038
f_{Λ_1}	Fraction of background with Λ_1 lifetime	0.718 ± 0.039
Λ_n	Inverse of lifetime of a background component	$0.040 \pm 0.032\,\mu$m^{-1}
Λ_1	Inverse of lifetime of a background component	$0.0439 \pm 0.0038\,\mu$m^{-1}
Λ_2	Inverse of lifetime of a background component	$0.0134 \pm 0.0010\,\mu$m^{-1}
a	Parameter in background fit to Φ	0.1441 ± 0.0062
b	Parameter in background fit to $\cos\Theta$	0.169 ± 0.013
$S_D(\text{OST})$	OST dilution scale factor	1.089 ± 0.049
$S_D(\text{SSKT})$	SSKT dilution scale factor	0.85 ± 0.18
$\epsilon_b(\text{OST})$	OST tagging efficiency for background	0.7592 ± 0.0019
$\epsilon_b(\text{SSKT})$	SSKT tagging efficiency for background	0.7266 ± 0.0026
$A(\text{OST})$	OST background tag asymmetry	0.4975 ± 0.0025
$A(\text{SSKT})$	SSKT background tag asymmetry	0.4956 ± 0.0033
$\epsilon_s(\text{OST})$	OST tagging efficiency for signal	0.9280 ± 0.0030
$\epsilon_s(\text{SSKT})$	SSKT tagging efficiency for signal	0.5222 ± 0.0067

Table 8.2 Summary of the final CDF $B_s^0 \to J/\psi\phi$ results for the physical observables from the fit with $\beta_s^{J/\psi\phi}$ fixed to its SM value

Observable	Complete sample
$\Delta\Gamma_s$ (ps^{-1})	0.068 ± 0.026
$\alpha_\perp$	0.279 ± 0.010
$\alpha_\parallel$	0.309 ± 0.012
$\delta_\perp$ (rad)	2.79 ± 0.53
$\delta_\parallel$ (rad)	3.09 ± 0.36
$c\tau_s$ (μm)	458.2 ± 5.8
A_{SW}	0.008 ± 0.018
δ_s (rad)	1.26 ± 0.75

Table 8.3 Matrix of relevant correlation coefficients among main parameters of SM analysis

	$\Delta\Gamma_s$	$\alpha_\perp$	$\alpha_\parallel$	$\delta_\perp$
$c\tau_s$	0.52	-0.16	0.07	0.03
$\Delta\Gamma_s$		-0.17	0.06	-0.01
$\alpha_\perp$			-0.53	-0.01
$\alpha_\parallel$				0.05

Table 8.4 Summary of the final CDF $B_s^0 \to J/\psi\phi$ fit results for the physical observables with $\beta_s^{J/\psi\phi}$ fixed to its SM value compared with the previous results [1]

Observable	5.2 fb^{-1} result	Complete sample
$\Delta\Gamma_s$ (ps^{-1})	0.075 ± 0.035	0.068 ± 0.026
$\alpha_\perp$	0.266 ± 0.014	0.279 ± 0.010
$\alpha_\parallel$	0.306 ± 0.015	0.309 ± 0.012
$\delta_\perp$ (rad)	2.95 ± 0.64	2.79 ± 0.53
$\delta_\parallel$ (rad)	3.08 ± 0.63	3.09 ± 0.36
$c\tau_s$ (μm)	458.6 ± 7.5	458.2 ± 5.8
A_{SW}	0.019 ± 0.027	0.008 ± 0.018
δ_s (rad)	1.37 ± 0.077	1.26 ± 0.75

Results for the SM fit using the full dataset and complete with systematic uncertainties are listed below

$$c\tau_s = 458.2 \pm 5.8 \ (stat) \pm 2.8 \ (syst) \ \mu\text{m},$$
$$\Delta\Gamma_s = 0.068 \pm 0.026 \ (stat) \pm 0.009 \ (syst) \ \text{ps}^{-1},$$
$$|A_\parallel|^2 = 0.229 \pm 0.010 \ (stat) \pm 0.014 \ (syst),$$
$$|A_0|^2 = 0.512 \pm 0.012 \ (stat) \pm 0.018 \ (syst),$$
$$\delta_\perp = 2.79 \pm 0.53 \ (stat) \pm 0.15 \ (syst).$$

Table 8.5 Comparison of the results obtained in this thesis (first column) with most recent results of other experiments [6–9], and the SM expectations in the rightmost column [5, 10]

Observable	CDF	D0	LHCb	ATLAS	SM
Signal yield	11 000	6 600	21 200	22 700	
$\Delta\Gamma_s$ (ps^{-1})	0.068 ± 0.027	$0.163^{+0.065}_{-0.064}$	0.116 ± 0.019	0.053 ± 0.023	0.087 ± 0.021
τ_s (ps)	1.528 ± 0.021	$1.443^{+0.038}_{-0.035}$	1.521 ± 0.020	1.477 ± 0.017	1.489 ± 0.031
$|A_0|^2$	0.512 ± 0.022	$0.558^{+0.017}_{-0.019}$	0.523 ± 0.025	0.528 ± 0.011	0.531 ± 0.022
$|A_\parallel|^2$	0.229 ± 0.017	$0.231^{+0.024}_{-0.030}$	0.255 ± 0.020	0.220 ± 0.011	0.230 ± 0.028
$\delta_\perp$	2.79 ± 0.55	–	2.90 ± 0.37	–	2.97 ± 0.18

The SM expectation of τ_s is given considering $\tau_s = (0.98 \pm 0.02)\tau_d$ [5], where τ_d is the world average value of the B^0 lifetime [11]. The theoretical predictions on the polarization amplitudes and $\delta_\perp$ are based on [12] and polarization measured in $B^0 \to J/\psi K^\star(892)^0$ decays [11]. In all experimental values, the uncertainties reported in the table includes in quadrature the statistical and the systematic uncertainty. The values of $\Delta\Gamma_s$, τ_s, $|A_0|^2$, and $|A_\parallel|^2$, from D0, LHCb, and ATLAS are obtained in the analysis with $\beta_s^{J/\psi\phi}$ not constrained to its SM value

All results are among the most precise determinations from a single experiment and exhibit an excellent agreement with the SM predictions and with measurements from other experiments (see Table 8.5). The value of $\Delta\Gamma_s$ agrees with the SM expectation, although the experimental precision does not completely rule out the occurrence of $\Delta\Gamma_s = 0$ yet. The polarization amplitudes are in good agreement with the measurements performed by Belle, Babar, and CDF [2–4] of the $B^0 \to J/\psi K^{*0}(892)$ amplitudes, confirming the validity of the $SU(3)$ approximate symmetry in this case. From the measurements of lifetime and decay width difference and their correlations we derive $\Delta\Gamma_s/\Gamma_s = 0.1045 \pm 0.048$ (*stat*) ± 0.027 (*syst*), while from the world average value of the B^0 lifetime we derive $\tau_s/\tau_d = 1.006 \pm 0.015$ (stat+syst). This last result supports the expectations from heavy quark effective theory models [5].

8.1.1 S-Wave Fraction

The fraction of S-wave determined by the fit in the KK mass range $[1.009 - 1.028]$ GeV/c^2 is 0.8 ± 1.6 %, consistent with zero. We set un upper limit on this fraction using the profile-likelihood ratio, described in detail in Sect. 8.2.1, yielding $A_{SW} < 5.6$ % at 95 % C.L. and $A_{SW} < 3.5$ % at 68 % C.L. (Fig. 8.4a). The vanishing value of observed S-wave fraction is consistent with results from a dedicated and more accurate study based on two-dimensional $m(J/\psi KK) - m(KK)$ fits (Appendix A). The S-wave results are incompatible with the findings of D0 on a similar mass range $(17.3 \pm 3.6$ %) but compatible with the more precise LHCb results [13].

Fig. 8.4 CL intervals for the fraction of S-wave (**a**)

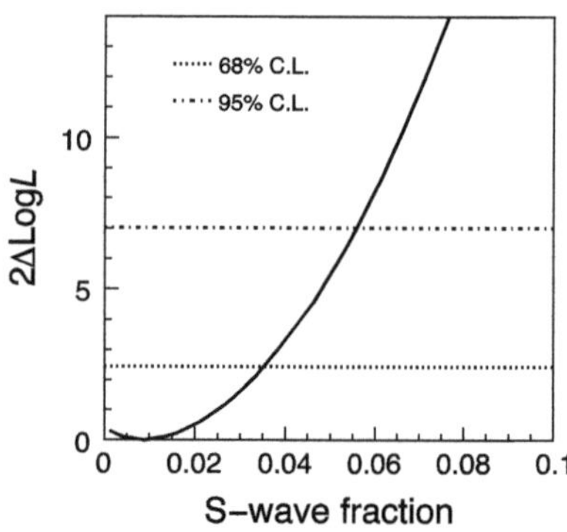

8.2 Mixing Phase

A further refinement of the estimation method is necessary to extract the main result of this thesis: the measurement of the CP-violating mixing phase, $\beta_s^{J/\psi\phi}$. Section 6.3 shows that non-negligible biases depending on the true values of the parameters affect the ML estimates of the full fit with $\beta_s^{J/\psi\phi}$ allowed to float, which makes extremely challenging and likely unreliable any attempt at bias corrections. In addition, the estimated uncertainties are unlikely to represent actual confidence intervals with the desired level of confidence.

Hence, an approach that is robust against the above effects is strongly desired to produce a reliable result for the mixing phase. Instead of quoting maximum likelihood estimates with standard uncertainties, results are given in the form of confidence intervals, in the one-dimensional space of $\beta_s^{J/\psi\phi}$ and the two-dimensional space of $(\beta_s^{J/\psi\phi}, \Delta\Gamma_s)$, by using the Neyman construction of a fully frequentist confidence region [14]. An arbitrariness is associated to the choice of the ordering algorithm, i.e., the procedure chosen to accumulate regions of the parameters space until the desired confidence level is attained. The choice for the ordering algorithm is discussed below.

8.2.1 The Profile-Likelihood Ratio

We construct the confidence regions using the profile likelihood ratio (PLR) ordering [15–17]. Constructing correct and informative confidence regions from highly multi-dimensional likelihoods is challenging. Determining the full 32-dimensional confidence space is computationally prohibitive. More importantly, the choice of the ordering algorithm has a non-trivial impact on the results. One needs to avoid that the projection of the fully-dimensional region onto the $(\beta_s^{J/\psi\phi}, \Delta\Gamma_s)$ subspace of interest includes most, if not all, of the allowed values, thus yielding a uninformative result. We choose to replace the likelihood, $\mathcal{L}(\beta_s^{J/\psi\phi}, \Delta\Gamma_s, \vartheta)$ with the profile likelihood, $\mathcal{L}_p(\beta_s^{J/\psi\phi}, \Delta\Gamma_s, \hat{\vartheta})$. For every point in the $(\beta_s^{J/\psi\phi}, \Delta\Gamma_s)$ plane, $\hat{\vartheta}$ are the values of nuisance parameters that maximize the likelihood. The profile-likelihood ratio $-2\Delta\ln(\mathcal{L}_p)$ is then used as a χ^2 variable to derive confidence regions in the two-dimensional space $(\beta_s^{J/\psi\phi}, \Delta\Gamma_s)$. However, simulations show that the observed distribution of $-2\Delta\ln(\mathcal{L}_p)$ deviates from the χ^2 distribution with two degrees of freedom $(\chi^2(2))$. Specifically, the resulting confidence regions contain true values of the parameters with lower probability than the nominal confidence level (C.L.) because the observed $-2\Delta\ln(\mathcal{L}_p)$ distribution has longer tails than a χ^2 distribution. In addition, the shape of the $-2\Delta\ln(\mathcal{L}_p)$ distribution in this problem appears to depend on the true values of the nuisance parameters, which are unknown. We therefore use the simulation of a large number of pseudoexperiments to derive the actual expected distribution of $-2\Delta\ln(\mathcal{L}_p)$. The effect of systematic uncertainties is accounted for by randomly sampling a limited number of points in the space of all nuisance parameters and using the most conservative of the resulting profile-likelihood ratio distributions to calculate the final confidence level.

We first fit the data with all parameters floating. Then, for each point in a 20×10 grid on the $(\beta_s^{J/\psi\phi}\Delta\Gamma_s)$ plane in the range $-\pi/2 < \beta_s^{J/\psi\phi} < \pi/2$ and $-0.3 < \Delta\Gamma_s < 0.3\,\mathrm{ps}^{-1}$, we fit the data by floating all parameters but $\beta_s^{J/\psi\phi}$ and $\Delta\Gamma_s$, which are fixed to the values corresponding to the probed point. Twice the negative difference between the logarithms of the likelihood values obtained in each of the two steps provide a profile-likelihood ratio value $-2\Delta\ln(\mathcal{L}_p)$ for each of the points in the $(\beta_s^{J/\psi\phi}\,\Delta\Gamma_s)$ plane. To derive confidence regions from the observed values of $-2\Delta\ln(\mathcal{L}_p)$, we compare them with the expected distribution of $-2\Delta\ln(\mathcal{L}_p)$. We obtain these distributions by generating for each $(\beta_s^{J/\psi\phi}\,\Delta\Gamma_s)$ value 16 ensembles of 1000 pseudoexperiments each. In each ensemble, the true values of $\beta_s^{J/\psi\phi}$ and $\Delta\Gamma_s$ correspond to the probed point, while the true values of the nuisance parameters are a random sampling from an hypercube centered at their best fit values in data, with side corresponding to ten standard deviations. Since the true values for these parameters are unknown, by sampling in a generous range of their physically-allowed values, we ensure that the resulting PLR distribution approximate what is likely to be the case in data. The size of the probed range of nuisance parameters is not arbitrary. Berger and Boos suggest that the C.L. range probed in the space of nuisance parameters should greatly exceed the C.L. chosen for quoting the final result [18]. Hence, since we quote 1σ and 2σ bands on the mixing phase and $\Delta\Gamma_s$, we choose to sample the nuisance parameters in a 5σ interval [18]. Profile-likelihood ratios are determined in each of these pseudoexperiment exactly as in data. The ensemble giving the broadest $-2\Delta\ln(\mathcal{L}_p)$ distribution is chosen. For each point in the $(\beta_s^{J/\psi\phi}, \Delta\Gamma_s)$ grid, we calculate the p-value as the fraction of pseudoexperiments from this ensemble in which a $-2\Delta\ln(\mathcal{L}_p)$ value as large or larger than in data is observed. The $(\beta_s^{J/\psi\phi}, \Delta\Gamma_s)$ region in which the p-value is larger than $1-$C.L. forms the C.L.% confidence region.

In practice we observe that the $-2\Delta\ln(\mathcal{L}_p)$ distribution is fairly independent of the value of $(\beta_s^{J/\psi\phi}, \Delta\Gamma_s)$ probed, so we do not need to generate pseudoexperiments for each ($\beta_s^{J/\psi\phi}$, $\Delta\Gamma_s$) pair. It suffices to compare the $-2\Delta\ln(L_p)$ observed in each ($\beta_s^{J/\psi\phi}$, $\Delta\Gamma_s$) point in data to just the expected $-2\Delta\ln(\mathcal{L}_p)$ distribution generated at a single arbitrary point. Because the chief goal of this analysis is to quantify compatibility of CDF data with the SM, we choose the SM point ($\beta_s^{J/\psi\phi} = 0.018$, $\Delta\Gamma_s = 0.087$) to generate the reference $-2\Delta\ln(\mathcal{L}_p)$ distribution. The deviations of the observed PLR distribution from the expected $\chi^2(2)$ distribution are the combined results of irregularities of the likelihood and the lack of accurate knowledge of nuisance parameters values, and amount to tipically 40 % of the nominal $-2\Delta\ln(\mathcal{L}_p)$ values.

 Because of the approximate symmetries of the likelihood, additional technical and numerical difficulties arise in the likelihood minimization required to construct the PLR, e.g., the fit may occasionally converge on a local minimum. Therefore, we determine two PLR values for each point of the plane. One is derived by starting the minimization in the $\Delta\Gamma > 0$ and $\delta_{\parallel} < \pi$ subplane, the other is computed starting the minimization in $\Delta\Gamma < 0$ and $\delta_{\parallel} > \pi$ plane, where 0 and π are symmetry points for those parameters. The simulation shows that this procedure allows accurate identification of the global minimum without imposing any constraint on the domain of fit parameters which would otherwise bias the results.

8.2.2 CP-Violating Results

Before proceeding with the construction of confidence regions, we inspect the fit projections when the $\beta_s^{J/\psi\phi}$ phase is left free to float in the fit. Figures 8.5 and 8.6 show the proper decay time and the transversity angles distributions of the candidates. These projections are similar to the one obtained for the fit with $\beta_s^{J/\psi\phi}$ fixed to the SM value, supporting once more the accuracy of the parametrization of the distribution of interest in the fit.

   CP-violating results are presented in Table 8.6. They are in agreement with the results of the CP-conserving fit in Table 8.1. The minimum of the likelihood is found to be at $\beta_s^{J/\psi\phi} = (0.11 \pm 0.12)$ and $\Delta\Gamma_s = (0.068 \pm 0.027)\,\text{ps}^{-1}$, in agreement with the SM expectation, although the central values could be biased and the uncertainties non Gaussian.

8.2.3 Confidence Contours

We present the confidence regions in the ($\beta_s^{J/\psi\phi}$, $\Delta\Gamma_s$) plane in Fig. 8.7 constructed using the profile-likelihood ratio ordering.

   The two-dimensional PLR distribution as function of $\beta_s^{J/\psi\phi}$ and $\Delta\Gamma_s$ is shown in Fig. 8.8, while the cumulative PLR distribution obtained with simulation is shown in Fig. 8.9a. The solid black line represents the PLR values for pseudoexperiments

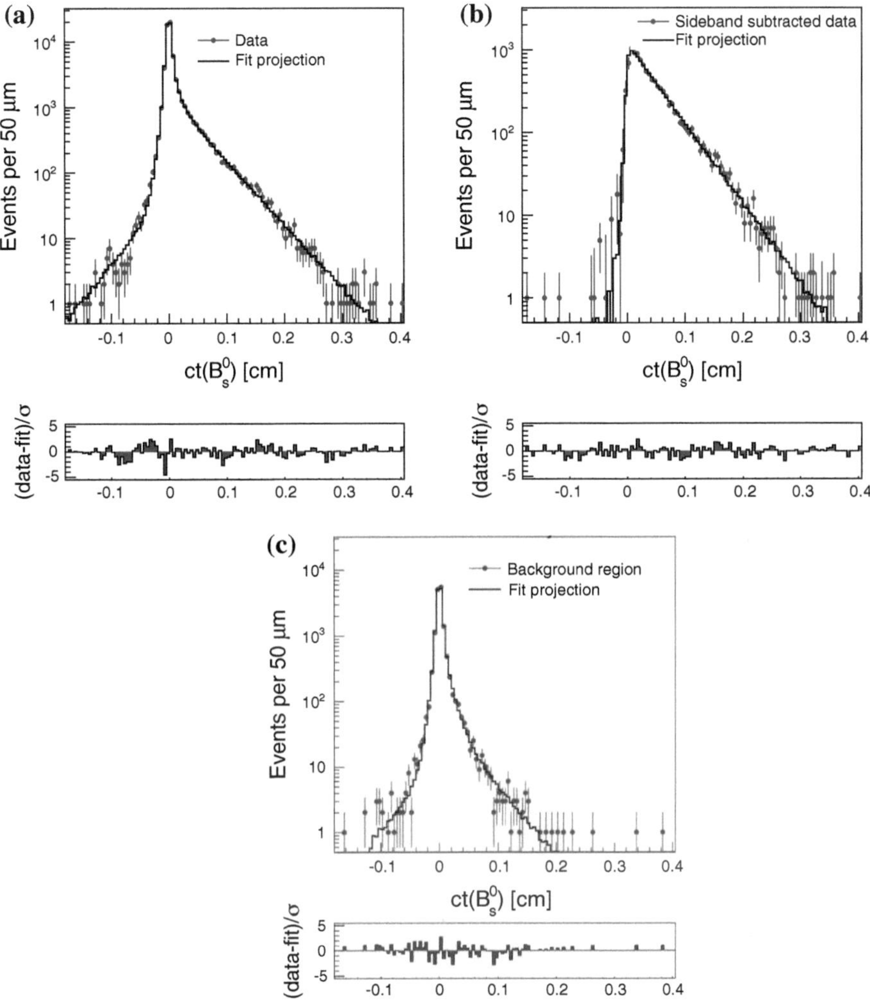

Fig. 8.5 Proper decay time distributions with fit projections overlaid when $\beta_s^{J/\psi\phi}$ is left free to float in the fit for all the events (**a**), and for events restricted to the $5.34 < m_{B_s^0} < 5.39\,\mathrm{GeV}/c^2$ region whose distribution is subtracted of sideband data distributions (**b**) and in the mass sideband region, with $(5.291 < m_{B_s^0} < 5.315\,\mathrm{GeV}/c^2) \cup (5.417 < m_{B_s^0} < 5.442\,\mathrm{GeV}/c^2)$ (**c**)

generated with the nuisance parameters set to the values observed in data. The colored dotted lines are PLR distributions for pseudoexperiments with varied nuisance parameters. The worst case is chosen to derive the final confidence regions. The distributions show that a PLR of 3.56 must be chosen to guarantee 68 % C.L. coverage, while a PLR of 8.13 corresponds to 95 % C.L. The adjusted confidence regions are reported in Fig. 8.10. The p-value corresponding to the pair of $(\beta_s^{J/\psi\phi},\ \Delta\Gamma_s)$ values

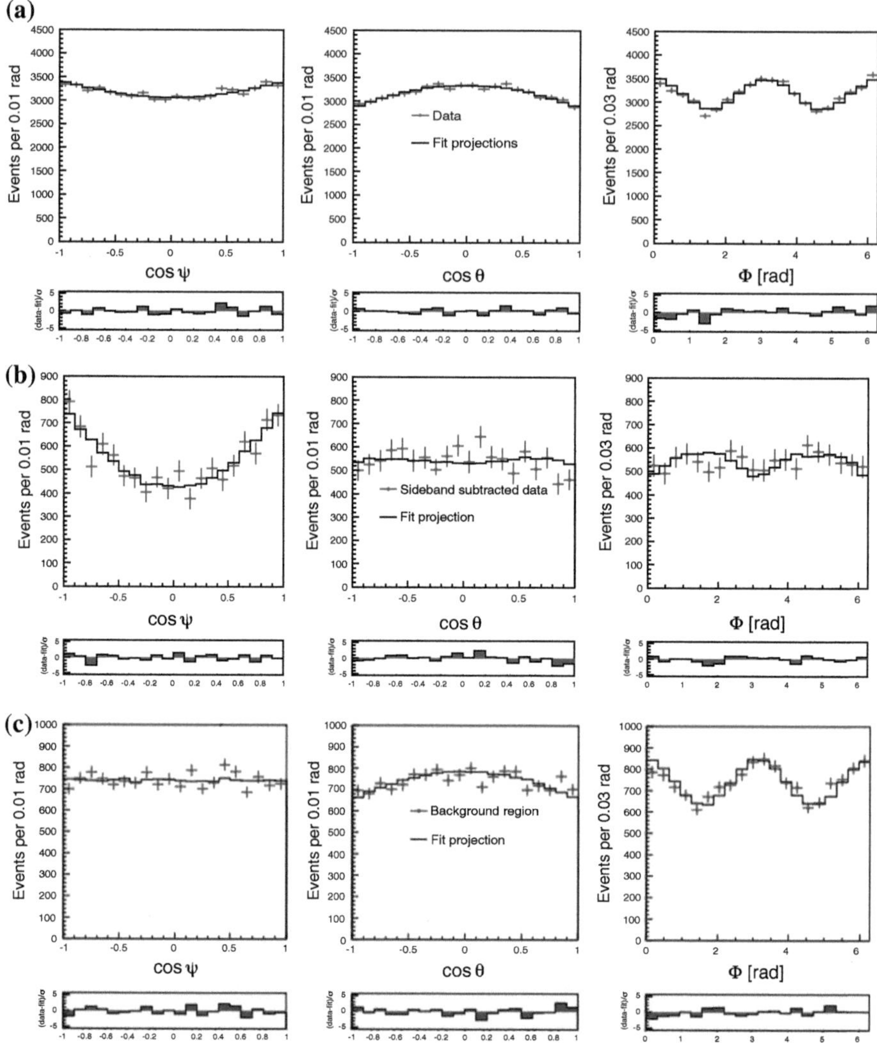

Fig. 8.6 Angular distributions with fit projection overlaid of the fit with $\beta_s^{J/\psi\phi}$ floating for the complete fit range (**a**), and for events restricted to the $5.34 < m_{B_s^0} < 5.39\,\mathrm{GeV}/c^2$ region whose distribution is subtracted of sideband data distributions (**b**) and in the mass sideband region, with $(5.291 < m_{B_s^0} < 5.315\,\mathrm{GeV}/c^2) \cup (5.417 < m_{B_s^0} < 5.442\,\mathrm{GeV}/c^2)$ (**c**)

predicted by the SM is 0.59 indicating full agreement of CDF data with the SM hypothesis.

Table 8.6 Results of the fit to mixing phase

Parameter	Description	Value
$\beta_s^{J/\psi\phi}$	CP-violating phase of B_s^0-$\bar{B}_s^0$ mixing amplitude	0.11 ± 0.12
$\Delta\Gamma_s$	Lifetime difference $\Gamma_L - \Gamma_H$	0.068 ± 0.027 ps^{-1}
$\alpha_\perp$	Fraction $|a_\perp|^2$	0.279 ± 0.011
$\alpha_\parallel$	Fraction $|a_\parallel|^2/(1 - |a_\perp|^2)$	0.309 ± 0.012
$\delta_\perp$	$Arg(A_\perp A_0^\star)$	2.73 ± 0.53
$\delta_\parallel$	$Arg(A_\parallel A_0^\star)$	3.04 ± 0.36
$c\tau$	B_s^0 mean lifetime	$457.9 \pm 5.8\,\mu$m
$|A_S|^2$	Fraction of S-wave K^+K^- component in the signal	0.008 ± 0.016
δ_S	Strong phase of S-wave amplitude	1.21 ± 0.65
Δm_s	B_s^0-$\bar{B}_s^0$ mixing frequency	17.72 ± 0.11 ps^{-1}
f_s	Signal fraction	0.1721 ± 0.0018
s_m	Mass uncertainty scale factor	1.727 ± 0.017
p_1	Mass background slope	-1.89 ± 0.42 (GeV/c^2)$^{-1}$
s_{ct_1}	Lifetime uncertainty scale factor 1	1.308 ± 0.012
s_{ct_2}	Lifetime uncertainty scale factor 2	3.34 ± 0.13
$f_\mathcal{R}$	Relative fraction of Gaussians of the decay-length resolution	0.852 ± 0.010
f_{pr}	Fraction of prompt background	0.8839 ± 0.0039
f_{Λ_n}	Fraction of background with Λ_n lifetime	0.210 ± 0.038
f_{Λ_1}	Fraction of background with Λ_1 lifetime	0.718 ± 0.039
Λ_n	Inverse of lifetime of a background component	$0.040 \pm 0.032\,\mu$m^{-1}
Λ_1	Inverse of lifetime of a background component	$0.0439 \pm 0.0038\,\mu$m^{-1}
Λ_2	Inverse of lifetime of a background component	$0.0134 \pm 0.0010\,\mu$m^{-1}
a	Parameter in background fit to Φ	0.1441 ± 0.0062
b	Parameter in background fit to $\cos\Theta$	0.169 ± 0.013
$S_D(\mathrm{OST})$	OST dilution scale factor	1.085 ± 0.049
$S_D(\mathrm{SSKT})$	SSKT dilution scale factor	0.87 ± 0.17
$\epsilon_b(\mathrm{OST})$	OST tagging efficiency for background	0.7592 ± 0.0019
$\epsilon_b(\mathrm{SSKT})$	SSKT tagging efficiency for background	0.7266 ± 0.0026
$A(\mathrm{OST})$	OST background tag asymmetry	0.4975 ± 0.0025
$A(\mathrm{SSKT})$	SSKT background tag asymmetry	0.4956 ± 0.0033
$\epsilon_s(\mathrm{OST})$	OST tagging efficiency for signal	0.9280 ± 0.0030
$\epsilon_s(\mathrm{SSKT})$	SSKT tagging efficiency for signal	0.5222 ± 0.0067

8.2.4 One-Dimensional Confidence Intervals on the Mixing Phase

To obtain the maximum possible information on the mixing phase, the PLR distribution is evaluated also as a function of $\beta_s^{J/\psi\phi}$ only, through the same method previously

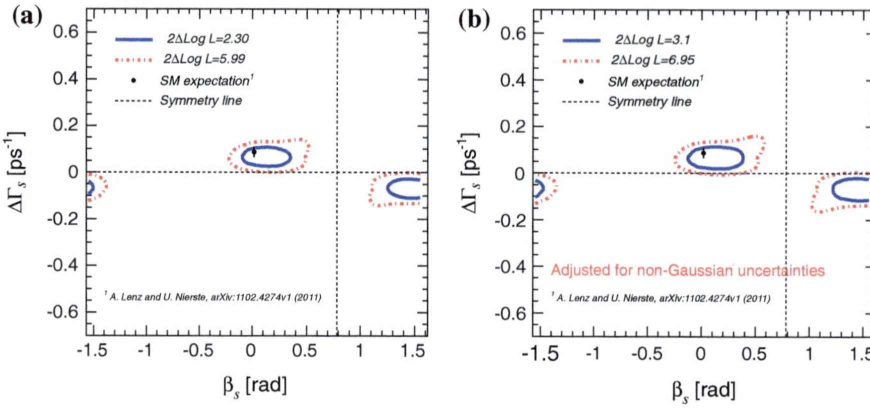

Fig. 8.7 *Curves* of iso-PLR in the $(\beta_s^{J/\psi\phi}, \Delta\Gamma_s)$ plane with no coverage adjustment (**a**) and adjusted for non-Gaussian uncertainties only (**b**)

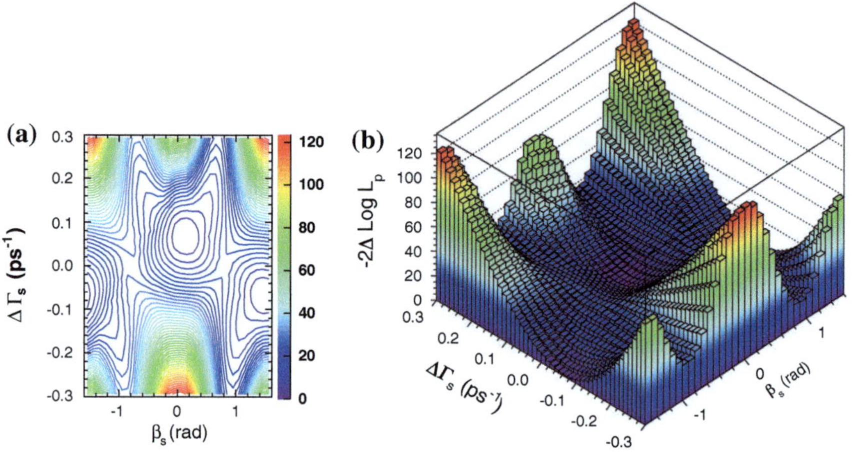

Fig. 8.8 Contours of iso-PLR (**a**) and two-dimensional distribution (**b**) in the $(\beta_s^{J/\psi\phi}, \Delta\Gamma_s)$ plane

described in which also $\Delta\Gamma_s$ is considered a nuisance parameter. We divide the interval $-\pi/2 < \beta_s^{J/\psi\phi} < \pi/2$ in 100 equally-spaced bins. We fit the data by floating all parameters but $\beta_s^{J/\psi\phi}$, which is fixed to the values corresponding to the point to probe, and calculate the corresponding PRL. Following a procedure analogous to of the two-dimensional case, we extract the correspondence between PLRs and p-values and use it to ensure proper coverage properties to the results (Fig. 8.9b). The iso-PLR curves corresponding to a 68 % C.L. and a 95 % C.L. interval are respectively 2.12 and 6.62. This correspond to $\beta_s^{J/\psi\phi} \in [-\pi/2, -1.51] \bigcup [-0.06, 0.30] \bigcup [1.26, \pi/2]$ at 68 % C.L., while $\beta_s^{J/\psi\phi} \in [-\pi/2, -1.36] \bigcup [-0.21, 0.53] \bigcup [1.04, \pi/2]$ at 95 %

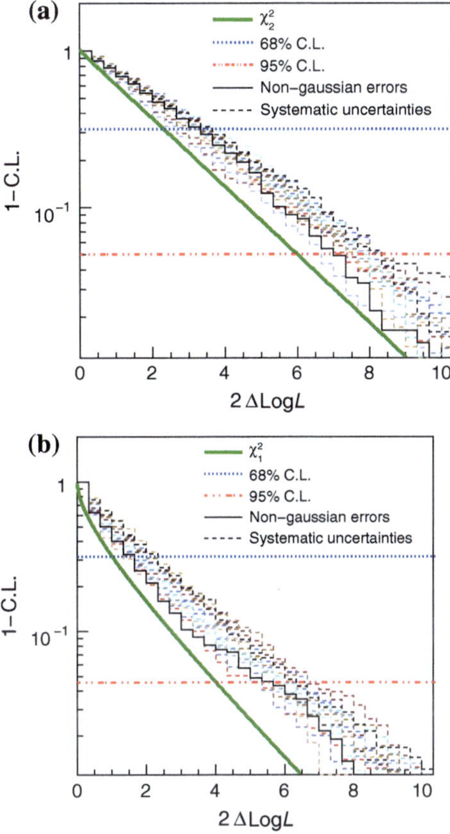

Fig. 8.9 **a** The *p*-value $(1 - \text{C.L.})$ as a function of two-dimensional PLR for the construction of the $(\beta_s^{J/\psi\phi}, \Delta\Gamma_s)$ confidence regions. The *green curve* is the cumulative of the χ^2 statistic corresponding to the ideal case. The *solid black* line represents PLR for pseudoexperiment generated with the nuisance parameters set to the values measured in data. The *colored dotted lines* are PLR distributions for pseudoexperiments that include systematic variations of nuisance parameters. The *p*-value as a function of one-dimensional PLR for the construction of the $\beta_s^{J/\psi\phi}$ confidence intervals (**b**)

C.L. (Fig. 8.11). The *p*-value with respect the SM is 0.54, indicating full agreement of CDF data with the SM hypothesis.

Table 8.7 compares the current and previous results. The current analysis achieves a 35 % improvement of the 68 % CL interval range and a marginal improvement on the 95 % CL interval range.

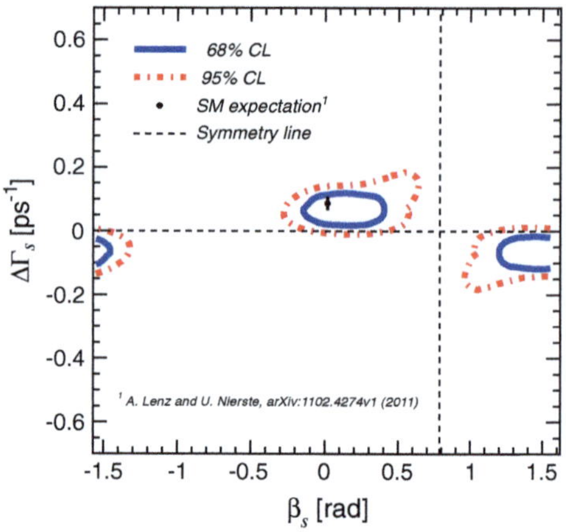

Fig. 8.10 Coverage-adjusted confidence regions in the $(\beta_s^{J/\psi\phi}, \Delta\Gamma_s)$ plane for the 68 % C.L. and 95 % C.L. This is the final result of the measurement

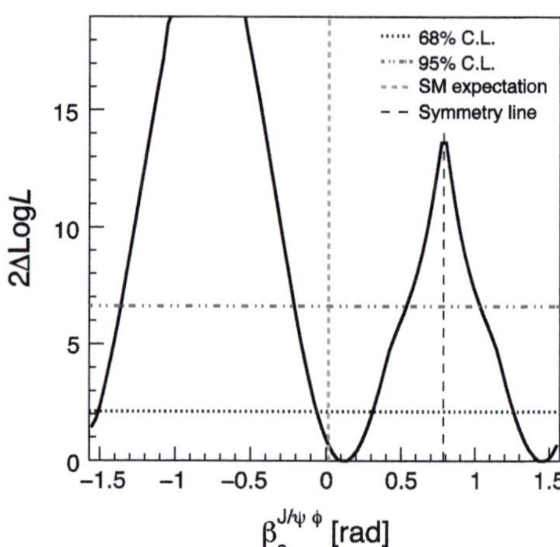

Fig. 8.11 Confidence regions in the $\beta_s^{J/\psi\phi}$ space

Table 8.7 Comparison of the new and previous intervals of $\beta_s^{J/\psi\phi}$ for the solution closest to the SM expectation

CL	5.2 fb⁻¹	9.6 fb⁻¹
68 %	[0.02, 0.52]	[−0.06, 0.30]
95 %	[−0.11, 0.65]	[−0.21, 0.53]
SM p-value	0.30	0.54

8.3 Consistency Checks

To check the consistency of the procedure, we first analyze only the first $5.2\,\mathrm{fb}^{-1}$ of data and check the compatibility with the previously-published CDF results to validate the fitting technique. Figure 8.12a shows the comparison iso-PRL curves before any coverage-corrections. The curves nearly overlap and the fit yields similar minima as expected [1]. Small residual discrepancies are likely due to fluctuations

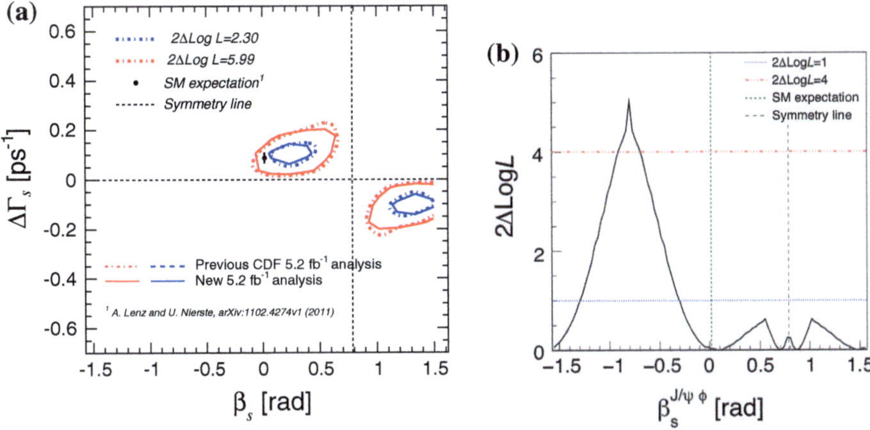

Fig. 8.12 Comparison of the iso-PLR *curves* as obtained in the previous CDF analysis (*dashed* contours) and the new analysis, both restricted to the first 5.2 fb^{-1} of the data (*solid* contours) (**a**). *Curves* of iso-PLR in the $\beta_s^{J/\psi\phi}$ plane (no coverage adjustment) from a fit to the newly added data only (last 4.4 fb^{-1}) where only OST is applied (**b**)

Table 8.8 Summary of the results of the main physics parameters in the fit of mixing phase using new data only (rightmost column)

Parameter	Previous result (5.2 fb^{-1})	New data only (4.4 fb^{-1})
$\beta_s^{J/\psi\phi}$	0.24 ± 0.13	0.09 ± 0.25
$\Delta\Gamma_s$ (ps^{-1})	0.097 ± 0.035	0.029 ± 0.033
$\alpha_\perp$	0.264 ± 0.014	0.297 ± 0.013
$\alpha_\parallel$	0.307 ± 0.015	0.309 ± 0.014
$\delta_\perp$	3.03 ± 0.52	3.50 ± 0.84
$\delta_\parallel$	3.02 ± 0.47	3.11 ± 0.40
$c\tau$ μm)	459.0 ± 7.3	456.4 ± 6.7

We report also the results of previous CDF measurement [1] (center column)

in the simulated samples used to derive them, use of slightly different tagging informations or offline requirements in event reconstruction. Figure 8.12b shows the one dimensional PLR profile for only the second half of the sample. The four minima due to the symmetries of the likelihood are evident and are related to the limited tagging power of the OST with respect to the SSKT, not used in last data, which does not suffice to completely lift the ambiguity due to the symmetry of the untagged likelihood. The results for the main physics parameters are consistent with those found in the first part of the sample (Table 8.8).

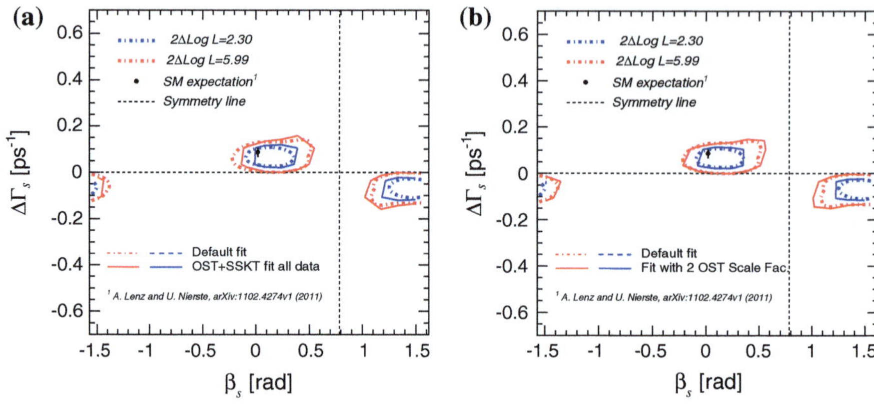

Fig. 8.13 Comparison of *curves* of iso-PLR in the $(\beta_s^{J/\psi\phi}, \Delta\Gamma_s)$ plane (no coverage adjustment) for the central fit and a fit in which SSKT is used in all data, assuming constant performances of the tagger in time (**a**). Comparison between the iso-PLR done for the default fit and the fit where two different OST scale factors are used for B_s^0 and for $\bar{B}_s^0$ (**b**)

Table 8.9 Comparison of the results obtained in this thesis (first column) with most recent results of other experiments (results of LHCb and ATLAS are preliminary) [6–9], and the SM expectations in the rightmost column [5, 10]

Quantity	CDF	D0	LHCb	ATLAS	SM
Signal yield	11,000	6,600	21,200	22,700	
$\beta_s^{J/\psi\phi}$	$[-0.06, 0.30]$ @68 % CL	$0.28^{+0.18}_{-0.19}$	0.001 ± 0.052	-0.11 ± 0.21	0.0184 ± 0.0009
$\Delta\Gamma_s$ (ps^{-1})	0.068 ± 0.027	$0.163^{+0.065}_{-0.064}$	0.116 ± 0.019	0.053 ± 0.023	0.087 ± 0.021
τ_s (ps)	1.528 ± 0.021	$1.443^{+0.038}_{-0.035}$	1.521 ± 0.020	1.477 ± 0.017	1.489 ± 0.031
$|A_0|^2$	0.512 ± 0.022	$0.558^{+0.017}_{-0.019}$	0.523 ± 0.025	0.528 ± 0.011	0.531 ± 0.022
$|A_\parallel|^2$	0.229 ± 0.017	$0.231^{+0.024}_{-0.030}$	0.255 ± 0.020	0.220 ± 0.011	0.230 ± 0.028
$\delta_\perp$	2.79 ± 0.55	–	2.90 ± 0.37	–	2.97 ± 0.18

The SM expectation of τ_s is given considering $\tau_s = (0.98 \pm 0.02)\tau_d$ [5], where τ_d is the world average value of the B^0 lifetime [11]. The theoretical predictions on the polarization amplitudes and $\delta_\perp$ are based on [12] and polarization measured in $B^0 \to J/\psi K^\star(892)^0$ decays [11]. The uncertainties of all experimental values include in quadrature statistical and systematic contributions. The values of $\beta_s^{J/\psi\phi}$ from D0 and ATLAS are obtained with constraints on the strong phases; in addition ATLAS use the constraint $\Delta\Gamma_s > 0$ as measured by LHCb. CDF does not use any constraint in the fit but choose to quote only the $\beta_s^{J/\psi\phi}$ interval corresponding to the $\Delta\Gamma_s > 0$ solution. The values of $\Delta\Gamma_s$, τ_s, $|A_0|^2$, and $|A_\parallel|^2$, from D0, LHCb and ATLAS are obtained in the analysis with $\beta_s^{J/\psi\phi}$ not constrained to its SM value

Figure 8.13a shows the comparison between the PLR distribution for the default fit and the fit where the SSKT is used in all data, assuming that the performance of the tagger is unchanged in recent data. The resulting confidence regions are very similar, confirming that our choice of not using an uncalibrated version of SSKT in the second half of data (Sect. 5.3.3) does not significantly compromise the quality of the result.

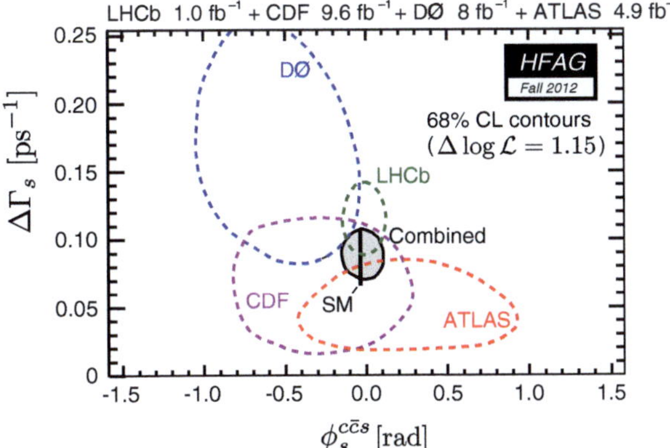

Fig. 8.14 The 68 % confidence level contours in the ($\phi_s^{c\bar{c}s}$, $\Delta\Gamma_s$) plane (where $\phi^{c\bar{c}s} = -2\beta_s^{J/\psi\phi}$) from CDF (this thesis), D0 [6], LHCb [7], and ATLAS [9], and their combined contour (*solid line* and *shaded area*), as well as the SM predictions (*black marker*) from global fit of the CKM matrix [5]. The combined result is consistent with the CKM predictions at the level of 0.14 standard deviations

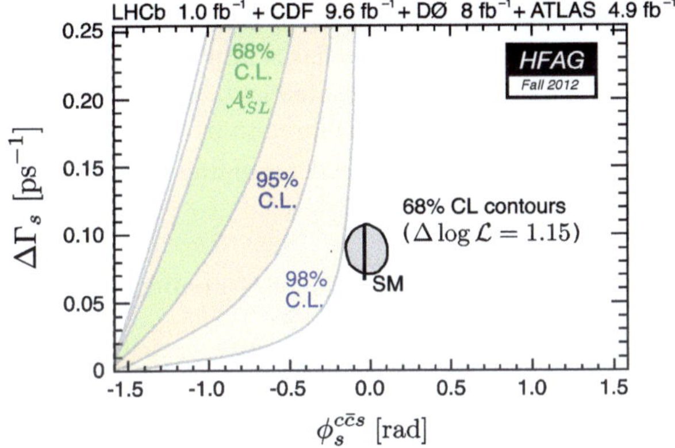

Fig. 8.15 The ($\phi^{c\bar{c}s}$, $\Delta\Gamma_s$) combined contour reported by HFAG [20] and the SM expectation together with the regions allowed at 68 and 95 % confidence level by the average measurements of $A_{sl} = -0.0109 \pm 0.0040$ [20] and $\Delta m_s = 17.719 \pm 0.043\,\mathrm{ps}^{-1}$ [20], through the relation $\tan\phi_s = A_{sl}\Delta m_s/\Delta\Gamma_s$; these regions are drawn assuming that any non-SM physics does not affect the phase difference $\phi^{c\bar{c}s} - \phi_s$

We also compute the PLR using two different OST scale factors, one for B_s^0 and the other for $\bar{B}_s^0$, to allow for a tagging asymmetry. This was the procedure adopted in the previous iteration of the analysis, since in the past calibration of the OST, the scale factors for the B^+ and the B^- mesons were found to differ by two standard deviations, contrary to the results from the more precise calibration reported in Sect. 5.3.2. The two scale factors are Gaussian-constrained in the fit to the values measured in the calibrations, $S^+ = 1.09 \pm 0.05$ and $S^- = 1.08 \pm 0.05$. The fit results are in excellent agreement results of the fit with a single scale factor. The resulting PLR is shown in Fig. 8.13b and compared with the PRL of the default fit. The confidence regions present a nearly perfect overlap.

Conclusions

Detailed studies of the weak interactions of quarks allow precise tests of the standard model and provide unique opportunities to search for new phenomena beyond the current theoretical knowledge. We report on a search for non-SM physics in the B_s^0 meson sector through the measurement of the decay-time-dependent CP-violating asymmetry of $B_s^0 \to J/\psi\,\phi$ decays using the full proton-antiproton collisions data set available to the CDF experiment and corresponding to approximately 10 fb^{-1}. The $B_s^0 \to J/\psi\phi\}$ decay is among the most sensitive probes of non-SM physics currently available in particle physics. It provides the best determination of the $\beta_s^{J/\psi\phi}$ CKM phase, which is reliably predicted to be small in the SM, potentially enhanced by a broad class of SM extensions, and poorly constrained experimentally.

The measurement is based on the analysis of decay-time dependent angular distributions of $B_s^0 \to J/\psi\phi\}$ decays whose flavor at production is known, exploiting an unbinned maximum likelihood fit. Detailed and extensive studies of the tagging algorithms resulted in a more accurate calibration of the opposite side tagger with an improvement of approximately 30 % in determining its performances. The $\beta_s^{J/\psi\phi}$ mixing phase is determined to be

$$-0.06 < \beta_s^{J/\psi\phi} < 0.30 \text{ at the 68 \% confidence level.}$$

Two-dimensional confidence intervals in the $(\beta_s^{J/\psi\phi}, \Delta\Gamma_s)$ plane are also reported. Precise measurements of the decay-width difference $\Delta\Gamma_s$; of the mean B_s^0 lifetime, τ_s; of the $B_s^0 \to J/\psi\phi\}$ polarization amplitudes, $|A_0|^2$ and $|A_\parallel|^2$, and of the CP-conserving phase $\delta_\perp$ are also provided:

$$\tau_s = 1.528 \pm 0.019\ (stat) \pm 0.009\ (syst)\ \text{ps},$$
$$\Delta\Gamma_s = 0.068 \pm 0.026\ (stat) \pm 0.009\ (syst)\ \text{ps}^{-1},$$
$$|A_\parallel|^2 = 0.229 \pm 0.010\ (stat) \pm 0.014\ (syst),$$
$$|A_0|^2 = 0.512 \pm 0.012\ (stat) \pm 0.018\ (syst),$$
$$\delta_\perp = 2.79 \pm 0.53\ (stat) \pm 0.15\ (syst).$$

All results are among the most precise determinations from a single experiment and exhibit an excellent agreement with the SM predictions and with measurements from other experiments (see Table 8.9). They are published in a Letter to *Physical Review* [19].

Figure 8.14 shows a comparison between the most recent experimental results in the $(\beta_s^{J/\psi\phi}, \Delta\Gamma_s)$ plane. The precision on the measurements of $\beta_s^{J/\psi\phi}$ and $\Delta\Gamma_s$ is still constrained by the limited samples sizes but is sufficient to indicate that no large deviations from the SM pattern are likely to occur. This is also confirmed by Fig. 8.15, which presents the combination [20] of all measurements again nicely consistent with the SM prediction. The model independent fit of all quark flavor measurements is consistent with the absence of large non-SM contributions in B_s^0-$\bar{B}_s^0$ mixing, although room for sizable deviations from the SM expectations is present. A discrepancy is evident between the direct determination of $\beta_s^{J/\psi\phi}$ and the indirect derivation through the dimuon asymmetry, A_{sl}, measured by D0 [21]. Assuming that non-SM physics in the B_s^0-$\bar{B}_s^0$ mixing is causing the D0 anomaly, the large observed dimuon asymmetry remains an unsolved puzzle, because it does not seem to be explainable by non-SM physics contributions to M_{12}^s alone. An enhancement of Γ_{12}^s compared to its SM value by a factor of $3 - 30$ would be required [5], which seems highly unlikely based on the large accumulated amount of results on tree-level B decays observed to be consistent with the SM. To investigate these issues further, independent, and more precise measurements of A_{sl} are needed. With this measurement CDF concludes a fruitful experimental program in the study of B_s^0 oscillations, which was pionereed in 2006 with their first observation, and in 2008 with the first measurement of the B_s^0 mixing phase.

References

1. T. Aaltonen et al. (CDF Collaboration), Measurement of the CP-violating phase $\beta_s^{J/\psi\phi}$ in $B_s^0{\to}J/\psi\phi$ decays with the CDF II detector. Phys. Rev. **D85**, 072002 (2012). arXiv:1112.1726 (hep-ex)
2. R. Itoh et al. (Belle Collaboration), Studies of CP violation in $B^0 \to J/\psi K^{*0}$ decays. Phys. Rev. Lett. **95**, 091601 (2005). arXiv:hep-ex/0504030 (hep-ex)
3. B. Aubert et al. (BaBar Collaboration), Measurement of decay amplitudes of $B \to (c\bar{c})K^*$ with an angular analysis, for $(c\bar{c}) = j/psi, /psi$ (2S) and chi(c1) (2006). arXiv:hep-ex/0607081 (hep-ex)
4. CDF Collaboration, Angular analysis of $B_s^0 \to J/\psi\phi$ and $B^0 \to J/\psi K^{*0}$ decays and measurement of $\Delta\Gamma_s$ and ϕ_s, CDF Note 8950 (2007)
5. A. Lenz, Theoretical update of B-mixing and lifetimes (2012). arXiv:1205.1444 (hep-ph)
6. V. Abazov et al. (D0 Collaboration), Measurement of the *CP* -violating phase $\phi_s^{J/\psi\phi}$ using the flavor-tagged decay $B_s^0{\to}J/\psi\phi$ in 8 fb^{-1} of $p\bar{p}$ collisions. Phys. Rev. **D85**, 032006 (2012). arXiv:1109.3166 (hep-ex)
7. The LHCb Collaboration, Tagged time-dependent angular analysis of $B_s^0{\to}J/\psi\phi$ decays at LHCb, LHCb-CONF-2012-002 (2012)
8. R. Aaij et al. (LHCb Collaboration), Measurement of the polarization amplitudes and triple product asymmetries in the $B_s^0{\to}\phi\phi$ decay. Phys. Lett. **B713**, 369 (2012). arXiv:1204.2813 (hep-ex)
9. G. Aad et al. (ATLAS Collaboration), Time-dependent angular analysis of the decay $B_s^0{\to}J/\psi\phi$ and extraction of $\Delta\Gamma_s$ and the *CP*-violating weak phase ϕ_s by ATLAS, (2012). arXiv:1208.0572 (hep-ex)
10. A. Datta et al., New physics in $b \to s$ transitions and the $B_{d,s}^0{\to}V_1 V_2$ angular analysis. Phys. Rev. **D86**, 076011 (2012). arXiv:1207.4495 (hep-ph)

11. J. Beringer et al., Particle data group, review of particle physics (RPP). Phys. Rev. **D86**, 010001 (2012)
12. M. Gronau, J.L. Rosner, Flavor symmetry for strong phases and determination of $\beta_s^{J/\psi\phi}$, $\Delta\Gamma_s$ in $B_s^0 \to J/\psi\phi$. Phys. Lett. **B669**, 321 (2008). arXiv:0808.3761 (hep-ph)
13. R. Aaij et al. (LHCb Collaboration), Measurement of the CP-violating phase ϕ_s in the decay $B_s^0 \to J/\psi\phi$. Phys. Rev. Lett. **108**, 101803 (2012). arXiv:1112.3183 (hep-ex)
14. J. Neyman, Outline of a theory of statistical estimation based on the classical theory of probability. Phil. Trans. R. Soc. A. **236**, 333–380 (1937)
15. G.J. Feldman, R.D. Cousins, A unified approach to the classical statistical analysis of small signals. Phys. Rev. **D57**, 3873 (1998). arXiv:physics/9711021 (physics.data-an)
16. K.S. Cranmer, Frequentist hypothesis testing with background uncertainty. eConf **C030908**, WEMT004 (2003). arXiv:physics/0310108 (physics)
17. G. Punzi, Ordering algorithms and confidence intervals in the presence of nuisance parameters. 88 (2005). arXiv:physics/0511202 (physics)
18. R.L. Berger, D.D. Boos, P values maximized over a confidence set for the nuisance parameter. J. Am. Stat. Assoc. **89**, 1012 (1994)
19. T. Aaltonen et al. (CDF Collaboration), Measurement of the bottom-strange meson mixing phase in the full CDF data set. Phys. Rev. Lett. **109**, 171802 (2012). arXiv:1208.2967 (hep-ex)
20. Y. Amhis et al. (Heavy Flavor Averaging Group), Averages of b-hadron, c-hadron, and tau-lepton properties as of early 2012 (2012). arXiv:1207.1158 (hep-ex), and updates at http://www.slac.stanford.edu/xorg/hfag
21. V.M. Abazov et al. (D0 Collaboration), Measurement of the anomalous like-sign dimuon charge asymmetry with 9 fb^{-1} of p pbar collisions. Phys. Rev. **D84**, 052007 (2011). arXiv:1106.6308 (hep-ex)

Appendix
Precise Determination of the S-Wave
Cross-Feed Background

A simpler, alternate fit has been used to check the determination of the size of the K^+K^- S-wave component and of the $B^0 \to J/\psi K\pi$ background. An auxiliary simultaneous likelihood fit of the unbinned $J/\psi K^+K^-$ and K^+K^- mass distributions is used to extract the fractions of interest, which are compared with the corresponding values obtained in the default fit. A few improvements are introduced over the simple K^+K^- mass fit described previously (Sect. 2.4.1) to better describe the details of the sample composition. The sample is enlarged by accepting events populating an extended K^+K^- mass window, which offers a longer lever arm to fit the features of the KK mass threshold. A slightly different selection with respect to the default analysis provides a greater sensitivity to the small S-wave component. We improve the templates of the $K\pi$ background in its full resonance structure, including both the P- and S- wave $K\pi$ along with their interference. The relative size of the $K\pi$ background is determined by the fit, we only fix its S-wave to P-wave ratio to the value measured by Babar [1]. With respect the default selection, the following changes are introduced:

- the KK mass window ranges from threshold to 1.2 GeV/c^2 (was 1.009–1.028) providing access to the S-wave-enriched low-mass region and high-mass data to properly model the $K^{\star 0}(892) \to K\pi$ background.
- the $ct(B) > 60$ μm requirement is introduced to suppress combinatorial background by preserving roughly 85 % of the signal.

The default threshold on the NN output (NN > 0.2) is applied. The $J/\psi K^+K^-$ mass distribution features three components, Fig. A.1a, a smooth, nearly-constant combinatorial background, which is modeled with a linear function; the B_s^0 signal component, which comprises both the P-wave ($J/\psi\phi$) and the S-wave ($J/\psi KK$) contributions, and a broad structure from $B^0 \to J/\psi K\pi$ decays in which the pion is misreconstructed as a kaon. The signal is modeled with a sum of three Gaussian functions whose parameters are extracted from a fit of a large sample of simulated decays and fixed. The B^0 line shape is determined from simulation.

© Springer International Publishing Switzerland 2015

S. Leo, *CP Violation in $B_s^0 \to J/\psi\phi$ Decays*, Springer Theses,

DOI 10.1007/978-3-319-07929-5

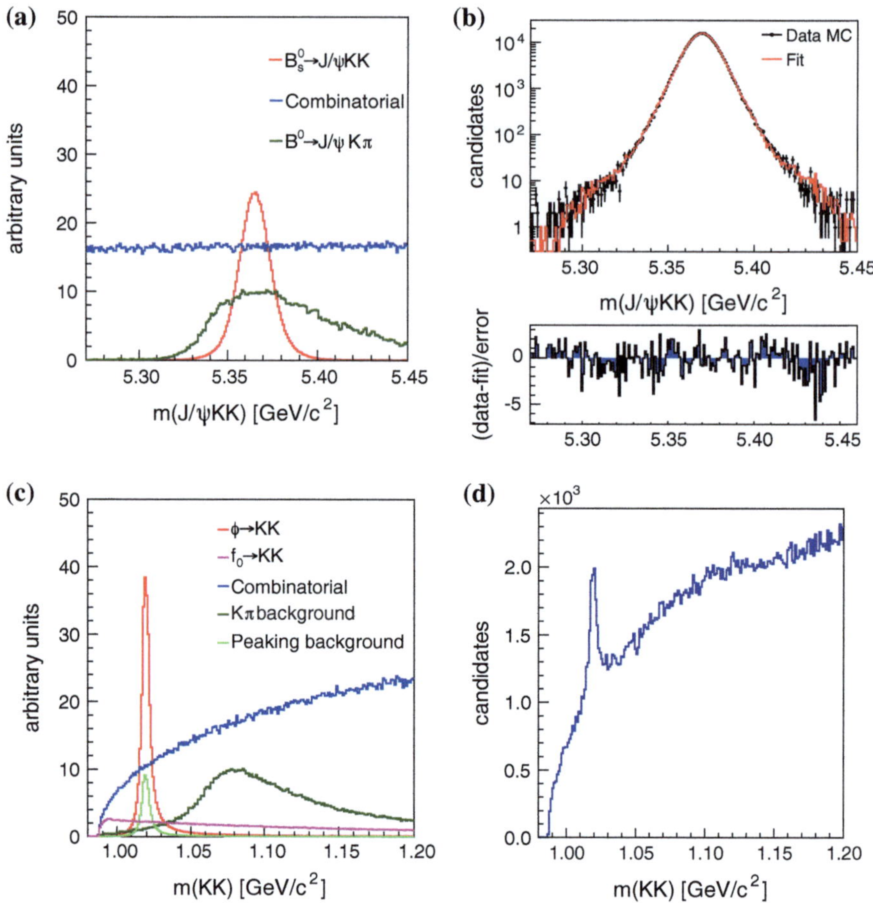

Fig. A.1 Simulated events (template) used to model the $J/\psi K^+ K^-$ spectrum (**a**). Distribution of $J/\psi K^+ K^-$ of simulated events with fit-projection overlaid to extract the B_s^0 signal template (**b**). Templates used to model the $K^+ K^-$ spectrum (**c**). Distribution of $K^+ K^-$ for events in the low-mass sideband of the B_s^0 peak with $ct(B) < 60\,\mu\mathrm{m}$ (**d**)

The $K^+ K^-$ mass distribution features five distinct components (Fig. A.1c): the narrow signal from ϕ mesons associated to a B_s^0 decay, the S-wave component, the $K\pi$ background due to misidentification of the pion as a kaon, the combinatorial background, and a peaking background presumably due to a real ϕ meson associated to a pair of random tracks or muons. The ϕ resonances are modeled with a relativistic Breit-Wigner with parameters fixed to those reported by the Particle Data Group [2], convoluted with a Gaussian resolution function with width determined by the fit. The S-wave component is modeled using a Flatté distribution [3] with parameters measured by BES [4]. The combinatorial background is empirically modeled from an histogram (Fig. A.1d) of the mass distribution of events populating the low-mass sideband of the B_s^0 peak, depleted in actual B^0 decays; such sample is selected by inverting the decay time requirement, $ct(B) < 60\,\mu$m, resulting in an independent sample. To adjust the normalization of the ϕ peaking background, an additional Breit-Wigner component is included in the background PDF, with fraction free to float, to account for its possible dependences on the B mass.

Because no simulation of the $B^0 \rightarrow J/\psi K\pi$ resonance structure is included in the standard generators, the $K\pi$ component is modeled from a custom simulation, where the S- and P-wave are simulated along with their interference as measured by Babar (Fig. A.1c). This simulation is validated on data by comparing the efficiency-unfolded angular distributions of simulated events and data and finding good agreement (Fig. A.2). The only additional external constraint is the $7.3 \pm 1.8\,\%$ ratio between S- and P-wave contributions with respect to a $K^\star(892)^0$ signal in the range $[0.8, 1.0]\,\mathrm{GeV}/c^2$ of the $K^+\pi^-$ spectrum [1].

The floating parameters are the overall fractions of B_s^0 and B^0 events, the fraction of S-wave events relative to the total B_s^0 events, the B_s^0 mass and width, the KK mass resolution, the slope of the combinatorial background in the B mass, and the fraction of peaking ϕ background. No fraction is constrained. The data distributions with fit projections are shown in Fig. A.3, showing good agreement. The determinations of the S-wave fraction and the B^0 fractions show some moderate sensitivity to the arbitrary modeling of the combinatorial background. With the current model, the S-wave contribution to the B_s^0 signal in the KK window used in the default fit is determined to be $(0.79 \pm 0.21)\,\%$, confirming the default fit result with much larger accuracy and previous CDF [5] and LHCb [6] determinations. The contribution of misreconstructed B^0 decays is found to be $(7.99 \pm 0.20)\,\%$ of the B_s^0 signal. This

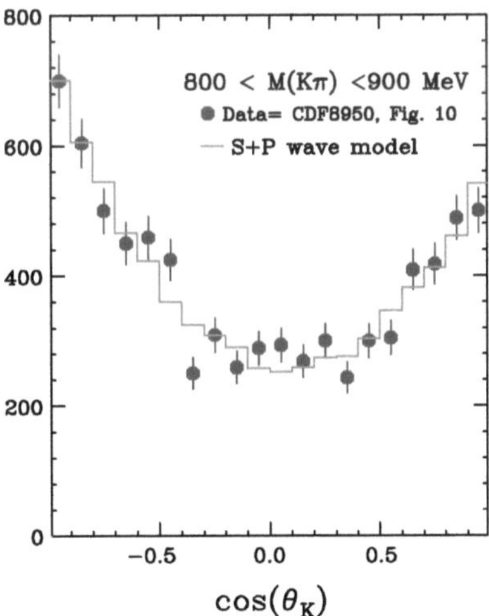

Fig. A.2 Comparison between the custom simulation and data for $B^0 \rightarrow J/\psi K\pi$ decays; the distribution is the transversity angle of the kaon, the most sensitive angular variable to the S–P interference; data are background subtracted and efficiency-unfolded

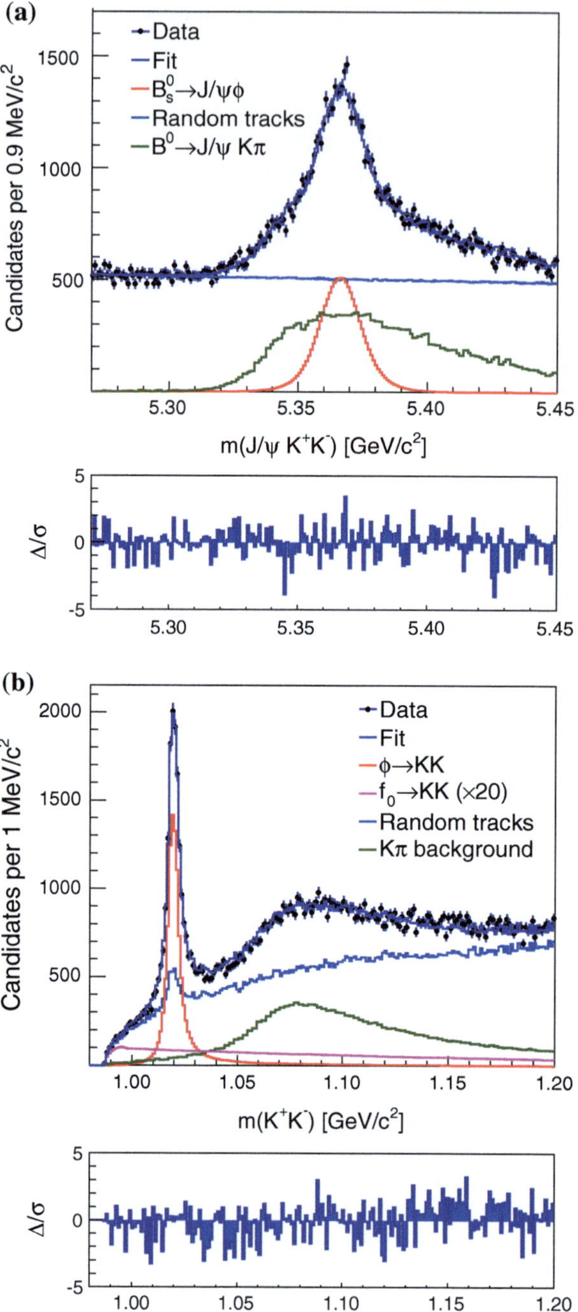

Fig. A.3 Distribution of $J/\psi K^+ K^-$ (**a**) and $K^+ K^-$ (**b**) masses with fit projections overlaid. The χ^2/d.o.f. are 251/200 and 336/194, respectively

is significantly higher than obtained by neglecting the resonant $B^0 \to J/\psi K\pi$ structure, and if overlooked could mimic an KK S-wave fraction larger than the one present in data. The present analysis is the first and only to date that pursue a detailed understanding of the $J/\psi K\pi$ contamination in the measurement of $\beta_s^{J/\psi\phi}$ using $B_s^0 \to J/\psi\phi$ decays.

References

1. B. Aubert et al., Ambiguity-free measurement of $\cos(2\beta)$: Timeintegrated and time-dependent angular analyses of $B \to J/\psi K\pi$. Phys. Rev. **D71**, 032005 (2005) arXiv:hep-ex/0411016 [hep-ex] (BaBar Collaboration)
2. J. Beringer et al., Review of particle physics (RPP). Phys. Rev. **D86**, 010001 (2012). (Particle Data Group).
3. S.M. Flatte, Coupled—channel analysis of the $\pi\eta$ and K anti-K systems near K anti-K threshold. Phys. Lett. **B63**, 224 (1976)
4. M. Ablikim et al., Resonances in $J/\psi \to \phi\pi^+\pi^-$ and $\phi K^+ K^-$. Phys. Lett. **B607**, 243 (2005) arXiv:hep-ex/0411001 [hep-ex] (BES Collaboration)
5. T. Aaltonen et al., Measurement of the CP-violating phase $\beta_s^{J/\psi\phi}$ in $B_s^0 \to J/\psi\phi$ decays with the CDF II detector. Phys. Rev. **D85**, 072002 (2012) arXiv:1112.1726 [hep-ex]
6. The LHCb Collaboration, Tagged time-dependent angular analysis of $B_s^0 \to J/\psi\phi$ decays at LHCb, LHCb-CONF-2012-002 (2012).